STUDENT'S SOL… MANUAL

SHARON MYERS
Radford University
KEYING YE
University of Texas at San Antonio

ESSENTIALS OF PROBABILITY & STATISTICS FOR ENGINEERS & SCIENTISTS

Ronald Walpole
Raymond Myers
Sharon Myers
Radford University
Keying Ye
University of Texas at San Antonio

PEARSON

Boston Columbus Indianapolis New York San Francisco Upper Saddle River
Amsterdam Cape Town Dubai London Madrid Milan Munich Paris Montreal Toronto
Delhi Mexico City Sao Paulo Sydney Hong Kong Seoul Singapore Taipei Tokyo

Reproduced by Pearson from electronic files supplied by the author.

Publishing as Pearson, 75 Arlington Street, Boston, MA 02116.

ISBN-13: 978-0-321-78399-8
ISBN-10: 0-321-78399-9

1 2 3 4 5 6 OPM 15 14 13 12 11

www.pearsonhighered.com

PEARSON

STUDENT'S SOLUTION MANUAL

KEYING YE AND SHARON MYERS

for

ESSENTIALS OF PROBABILITY & STATISTICS FOR ENGINEERS & SCIENTISTS

Solutions of Odd-Numbered Non-Review Exercise Problems

WALPOLE, MYERS, MYERS, YE

Contents

Chapter 1

Introduction to Statistics and Probability

1.1 (a) $S = \{8, 16, 24, 32, 40, 48\}$.

(b) For $x^2 + 4x - 5 = (x+5)(x-1) = 0$, the only solutions are $x = -5$ and $x = 1$. $S = \{-5, 1\}$.

(c) $S = \{T, HT, HHT, HHH\}$.

(d) $S = \{\text{N. America, S. America, Europe, Asia, Africa, Australia, Antarctica}\}$.

(e) Solving $2x - 4 \geq 0$ gives $x \geq 2$. Since we must also have $x < 1$, it follows that $S = \phi$.

1.3 (a) $A = \{1, 3\}$.

(b) $B = \{1, 2, 3, 4, 5, 6\}$.

(c) $C = \{x \mid x^2 - 4x + 3 = 0\} = \{x \mid (x-1)(x-3) = 0\} = \{1, 3\}$.

(d) $D = \{0, 1, 2, 3, 4, 5, 6\}$. Clearly, $A = C$.

1.5 $S = \{1HH, 1HT, 1TH, 1TT, 2H, 2T, 3HH, 3HT, 3TH, 3TT, 4H, 4T, 5HH, 5HT, 5TH, 5TT, 6H, 6T\}$.

1.7 (a) $S = \{M_1M_2, M_1F_1, M_1F_2, M_2M_1, M_2F_1, M_2F_2, F_1M_1, F_1M_2, F_1F_2, F_2M_1, F_2M_2, F_2F_1\}$.

(b) $A = \{M_1M_2, M_1F_1, M_1F_2, M_2M_1, M_2F_1, M_2F_2\}$.

(c) $B = \{M_1F_1, M_1F_2, M_2F_1, M_2F_2, F_1M_1, F_1M_2, F_2M_1, F_2M_2\}$.

(d) $C = \{F_1F_2, F_2F_1\}$.

(e) $A \cap B = \{M_1F_1, M_1F_2, M_2F_1, M_2F_2\}$.

(f) $A \cup C = \{M_1M_2, M_1F_1, M_1F_2, M_2M_1, M_2F_1, M_2F_2, F_1F_2, F_2F_1\}$.

(g)
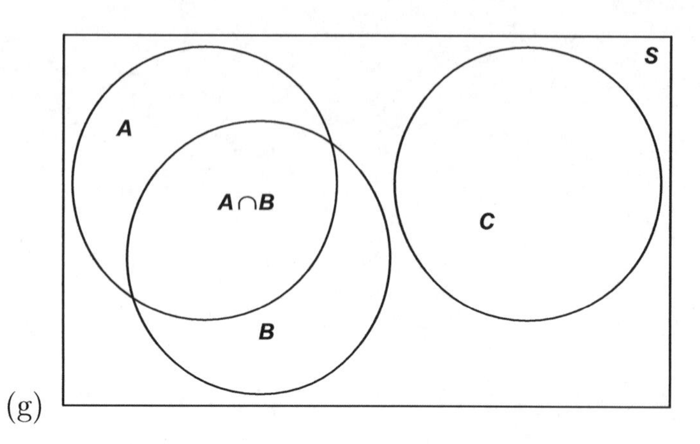

1.9 A Venn diagram is shown next.

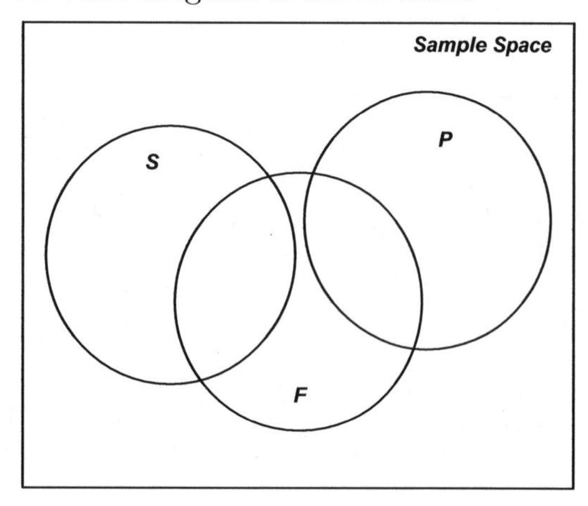

1.11 (a) $A \cup C = \{0, 2, 3, 4, 5, 6, 8\}$.

(b) $A \cap B = \phi$.

(c) $C' = \{0, 1, 6, 7, 8, 9\}$.

(d) $C' \cap D = \{1, 6, 7\}$, so $(C' \cap D) \cup B = \{1, 3, 5, 6, 7, 9\}$.

(e) $(S \cap C)' = C' = \{0, 1, 6, 7, 8, 9\}$.

(f) $A \cap C = \{2, 4\}$, so $A \cap C \cap D' = \{2, 4\}$.

1.13 A Venn diagram is shown next.

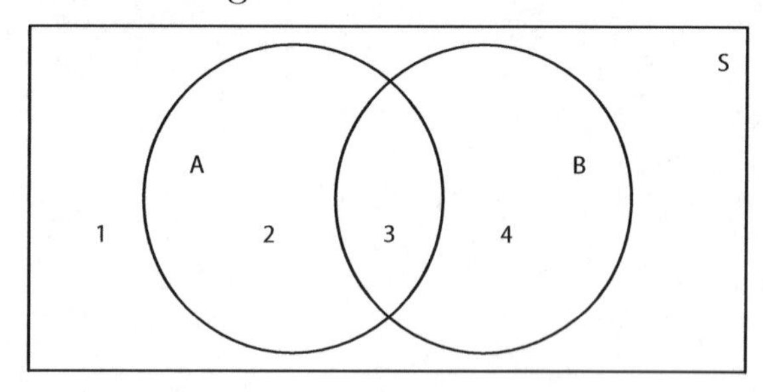

(a) From the above Venn diagram, $(A \cap B)'$ contains the regions of 1, 2 and 4.

(b) $(A \cup B)'$ contains region 1.

(c) A Venn diagram is shown next.

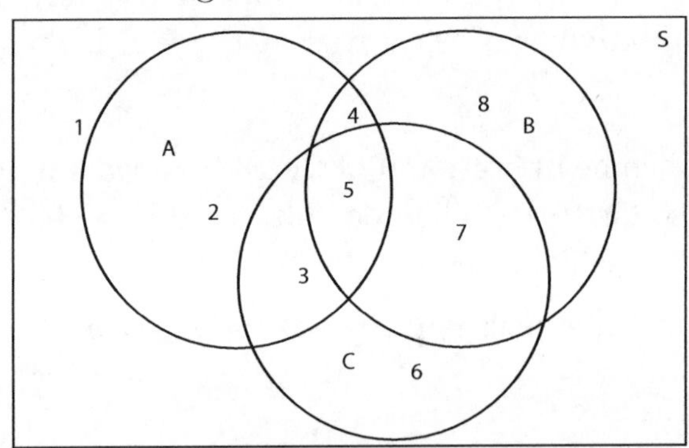

$(A \cap C) \cup B$ contains the regions of 3, 4, 5, 7 and 8.

1.15 (a) The family will experience mechanical problems but will receive no ticket for traffic violation and will not arrive at a campsite that has no vacancies.

(b) The family will receive a traffic ticket and arrive at a campsite that has no vacancies but will not experience mechanical problems.

(c) The family will experience mechanical problems and will arrive at a campsite that has no vacancies.

(d) The family will receive a traffic ticket but will not arrive at a campsite that has no vacancies.

(e) The family will not experience mechanical problems.

1.17 With $n_1 = 6$ sightseeing tours each available on $n_2 = 3$ different days, the multiplication rule gives $n_1 n_2 = (6)(3) = 18$ ways for a person to arrange a tour.

1.19 Since a student may be classified according to $n_1 = 4$ class standing and $n_2 = 2$ gender classifications, the multiplication rule gives $n_1 n_2 = (4)(2) = 8$ possible classifications for the students.

1.21 Using the generalized multiplication rule, there are $n_1 \times n_2 \times n_3 \times n_4 = (4)(3)(2)(2) = 48$ different house plans available.

1.23 With $n_1 = 3$ race cars, $n_2 = 5$ brands of gasoline, $n_3 = 7$ test sites, and $n_4 = 2$ drivers, the generalized multiplication rule yields $(3)(5)(7)(2) = 210$ test runs.

1.25 Since the first digit is a 5, there are $n_1 = 9$ possibilities for the second digit and then $n_2 = 8$ possibilities for the third digit. Therefore, by the multiplication rule there are $n_1 n_2 = (9)(8) = 72$ registrations to be checked.

1.27 The first house can be placed on any of the $n_1 = 9$ lots, the second house on any of the remaining $n_2 = 8$ lots, and so forth. Therefore, there are $9! = 362,880$ ways to place the 9 homes on the 9 lots.

1.29 The first seat must be filled by any of 5 girls and the second seat by any of 4 boys. Continuing in this manner, the total number of ways to seat the 5 girls and 4 boys is $(5)(4)(4)(3)(3)(2)(2)(1)(1) = 2880$.

1.31 (a) Any of the $n_1 = 8$ finalists may come in first, and of the $n_2 = 7$ remaining finalists can then come in second, and so forth. By Theorem 2.3, there $8! = 40320$ possible orders in which 8 finalists may finish the spelling bee.

(b) The possible orders for the first three positions are ${}_8P_3 = \frac{8!}{5!} = 336$.

1.33 By Theorem 2.4, ${}_6P_4 = \frac{6!}{2!} = 360$.

1.35 By Theorem 2.5, there are $4! = 24$ ways.

1.37 Assume February 29th as March 1st for the leap year. There are total 365 days in a year. The number of ways that all these 60 students will have different birth dates (i.e, arranging 60 from 365) is ${}_{365}P_{60}$. This is a very large number.

1.39 (a) Sum of the probabilities exceeds 1.

(b) Sum of the probabilities is less than 1.

(c) A negative probability.

(d) Probability of both a heart and a black card is zero.

1.41 Consider the events
S: industry will locate in Shanghai,
B: industry will locate in Beijing.

(a) $P(S \cap B) = P(S) + P(B) - P(S \cup B) = 0.7 + 0.4 - 0.8 = 0.3$.

(b) $P(S' \cap B') = 1 - P(S \cup B) = 1 - 0.8 = 0.2$.

1.43 $S = \{\$10, \$25, \$100\}$ with weights $275/500 = 11/20$, $150/500 = 3/10$, and $75/500 = 3/20$, respectively. The probability that the first envelope purchased contains less than \$100 is equal to $11/20 + 3/10 = 17/20$.

1.45 (a) $P(M \cup H) = 88/100 = 22/25$;

(b) $P(M' \cap H') = 12/100 = 3/25$;

(c) $P(H \cap M') = 34/100 = 17/50$.

1.47 (a) 0.32;

(b) 0.68;

(c) office or den.

1.49 $P(A) = 0.2$ and $P(B) = 0.35$

(a) $P(A') = 1 - 0.2 = 0.8$;

(b) $P(A' \cap B') = 1 - P(A \cup B) = 1 - 0.2 - 0.35 = 0.45$;

(c) $P(A \cup B) = 0.2 + 0.35 = 0.55$.

1.51 (a) $0.12 + 0.19 = 0.31$;

(b) $1 - 0.07 = 0.93$;

(c) $0.12 + 0.19 = 0.31$.

1.53 (a) $P(C) = 1 - P(A) - P(B) = 1 - 0.990 - 0.001 = 0.009$;

(b) $P(B') = 1 - P(B) = 1 - 0.001 = 0.999$;

(c) $P(B) + P(C) = 0.01$.

1.55 (a) $1 - 0.95 - 0.002 = 0.048$;

(b) $(\$25.00 - \$20.00) \times 10,000 = \$50,000$;

(c) $(0.05)(10,000) \times \$5.00 + (0.05)(10,000) \times \$20 = \$12,500$.

1.57 (a) The probability that a convict who pushed dope, also committed armed robbery.

(b) The probability that a convict who committed armed robbery, did not push dope.

(c) The probability that a convict who did not push dope also did not commit armed robbery.

1.59 (a) 0.018;

(b) $0.22 + 0.002 + 0.160 + 0.102 + 0.046 + 0.084 = 0.614$;

(c) $0.102/0.614 = 0.166$;

(d) $\frac{0.102+0.046}{0.175+0.134} = 0.479$.

1.61 Consider the events:
A: the vehicle is a camper,
B: the vehicle has Canadian license plates.

(a) $P(B \mid A) = \frac{P(A \cap B)}{P(A)} = \frac{0.09}{0.28} = \frac{9}{28}$.

(b) $P(A \mid B) = \frac{P(A \cap B)}{P(B)} = \frac{0.09}{0.12} = \frac{3}{4}$.

(c) $P(B' \cup A') = 1 - P(A \cap B) = 1 - 0.09 = 0.91$.

1.63 Consider the events:
A: the doctor makes a correct diagnosis,
B: the patient sues.
$P(A' \cap B) = P(A')P(B \mid A') = (0.3)(0.9) = 0.27$.

1.65 (a) 0.43;

(b) $(0.53)(0.22) = 0.12$;

(c) $1 - (0.47)(0.22) = 0.90$.

1.67 Let A and B represent the availability of each fire engine.

(a) $P(A' \cap B') = P(A')P(B') = (0.04)(0.04) = 0.0016.$

(b) $P(A \cup B) = 1 - P(A' \cap B') = 1 - 0.0016 = 0.9984.$

1.69 This is a parallel system of two series subsystems.

(a) $P = 1 - [1 - (0.7)(0.7)][1 - (0.8)(0.8)(0.8)] = 0.75112.$

(b) $P = \frac{P(A' \cap C \cap D \cap E)}{P\text{system works}} = \frac{(0.3)(0.8)(0.8)(0.8)}{0.75112} = 0.2045.$

1.71 Define S: the system works.
$P(A' \mid S') = \frac{P(A' \cap S')}{P(S')} = \frac{P(A')(1-P(C \cap D \cap E))}{1-P(S)} = \frac{(0.3)[1-(0.8)(0.8)(0.8)]}{1-0.75112} = 0.588.$

1.73 Consider the events:
C: an adult selected has cancer,
D: the adult is diagnosed as having cancer.
$P(C) = 0.05$, $P(D \mid C) = 0.78$, $P(C') = 0.95$ and $P(D \mid C') = 0.06$. So, $P(D) = P(C \cap D) + P(C' \cap D) = (0.05)(0.78) + (0.95)(0.06) = 0.096.$

1.75 $P(C \mid D) = \frac{P(C \cap D)}{P(D)} = \frac{0.039}{0.096} = 0.40625.$

1.77 Consider the events:
A: no expiration date,
B_1: John is the inspector, $P(B_1) = 0.20$ and $P(A \mid B_1) = 0.005$,
B_2: Tom is the inspector, $P(B_2) = 0.60$ and $P(A \mid B_2) = 0.010$,
B_3: Jeff is the inspector, $P(B_3) = 0.15$ and $P(A \mid B_3) = 0.011$,
B_4: Pat is the inspector, $P(B_4) = 0.05$ and $P(A \mid B_4) = 0.005$,
$P(B_1 \mid A) = \frac{(0.005)(0.20)}{(0.005)(0.20)+(0.010)(0.60)+(0.011)(0.15)+(0.005)(0.05)} = 0.1124.$

1.79 Consider the events:
A: a customer purchases latex paint,
A': a customer purchases semigloss paint,
B: a customer purchases rollers.
$P(A \mid B) = \frac{P(B \mid A)P(A)}{P(B \mid A)P(A)+P(B \mid A')P(A')} = \frac{(0.60)(0.75)}{(0.60)(0.75)+(0.25)(0.30)} = 0.857.$

Chapter 2

Random Variables, Distributions, and Expectations

2.1 Discrete; continuous; continuous; discrete; discrete; continuous.

2.3 A table of sample space and assigned values of the random variable is shown next.

Sample Space	w
HHH	3
HHT	1
HTH	1
THH	1
HTT	-1
THT	-1
TTH	-1
TTT	-3

2.5 (a) $c = 1/30$ since $1 = \sum_{x=0}^{3} c(x^2 + 4) = 30c$.

(b) $c = 1/10$ since

$$1 = \sum_{x=0}^{2} c\binom{2}{x}\binom{3}{3-x} = c\left[\binom{2}{0}\binom{3}{3} + \binom{2}{1}\binom{3}{2} + \binom{2}{2}\binom{3}{1}\right] = 10c.$$

2.7 (a) $P(X < 1.2) = \int_0^1 x\,dx + \int_1^{1.2}(2-x)\,dx = \left.\frac{x^2}{2}\right|_0^1 + \left.\left(2x - \frac{x^2}{2}\right)\right|_1^{1.2} = 0.68.$

(b) $P(0.5 < X < 1) = \int_{0.5}^1 x\,dx = \left.\frac{x^2}{2}\right|_{0.5}^1 = 0.375.$

2.9 We can select x defective sets from 2, and $3-x$ good sets from 5 in $\binom{2}{x}\binom{5}{3-x}$ ways. A random selection of 3 from 7 sets can be made in $\binom{7}{3}$ ways. Therefore,

$$f(x) = \frac{\binom{2}{x}\binom{5}{3-x}}{\binom{7}{3}}, \qquad x = 0, 1, 2.$$

In tabular form

x	0	1	2
$f(x)$	2/7	4/7	1/7

The following is a probability histogram:

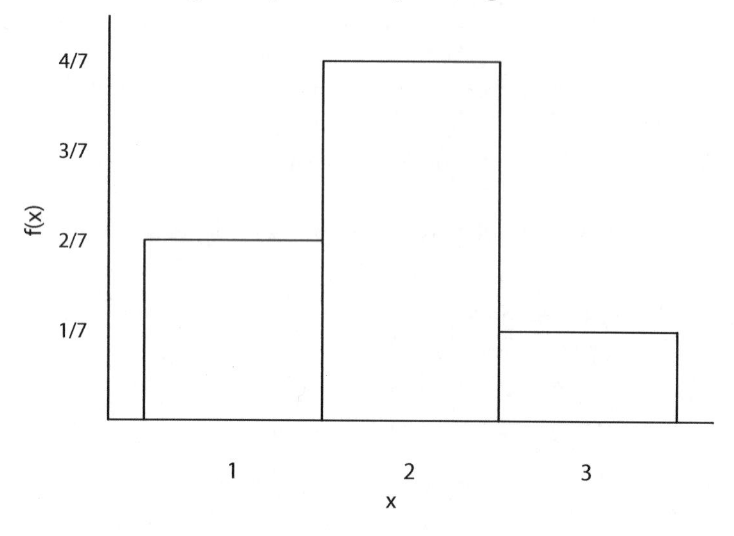

2.11 The c.d.f. of X is

$$F(x) = \begin{cases} 0, & \text{for } x < 0, \\ 0.41, & \text{for } 0 \le x < 1, \\ 0.78, & \text{for } 1 \le x < 2, \\ 0.94, & \text{for } 2 \le x < 3, \\ 0.99, & \text{for } 3 \le x < 4, \\ 1, & \text{for } x \ge 4. \end{cases}$$

2.13 The c.d.f. of X is

$$F(x) = \begin{cases} 0, & \text{for } x < 0, \\ 2/7, & \text{for } 0 \le x < 1, \\ 6/7, & \text{for } 1 \le x < 2, \\ 1, & \text{for } x \ge 2. \end{cases}$$

(a) $P(X = 1) = P(X \le 1) - P(X \le 0) = 6/7 - 2/7 = 4/7$;

(b) $P(0 < X \le 2) = P(X \le 2) - P(X \le 0) = 1 - 2/7 = 5/7$.

2.15 (a) $1 = k\int_0^1 \sqrt{x}\, dx = \frac{2k}{3}x^{3/2}\big|_0^1 = \frac{2k}{3}$. Therefore, $k = \frac{3}{2}$.

(b) For $0 \le x < 1$, $F(x) = \frac{3}{2}\int_0^x \sqrt{t}\, dt = t^{3/2}\big|_0^x = x^{3/2}$.
Hence,

$$F(x) = \begin{cases} 0, & x < 0 \\ x^{3/2}, & 0 \le x < 1 \\ 1, & x \ge 1 \end{cases}$$

$P(0.3 < X < 0.6) = F(0.6) - F(0.3) = (0.6)^{3/2} - (0.3)^{3/2} = 0.3004$.

2.17 Let T be the total value of the three coins. Let D and N stand for a dime and nickel, respectively. Since we are selecting without replacement, the sample space containing elements for which $t = 20, 25$, and 30 cents corresponding to the selecting of 2 nickels and 1 dime, 1 nickel and 2 dimes, and 3 dimes. Therefore, $P(T = 20) = \frac{\binom{2}{2}\binom{4}{1}}{\binom{6}{3}} = \frac{1}{5}$,
$P(T = 25) = \frac{\binom{2}{1}\binom{4}{2}}{\binom{6}{3}} = \frac{3}{5}$,
$P(T = 30) = \frac{\binom{4}{3}}{\binom{6}{3}} = \frac{1}{5}$,
and the probability distribution in tabular form is

t	20	25	30
$P(T = t)$	1/5	3/5	1/5

As a probability histogram

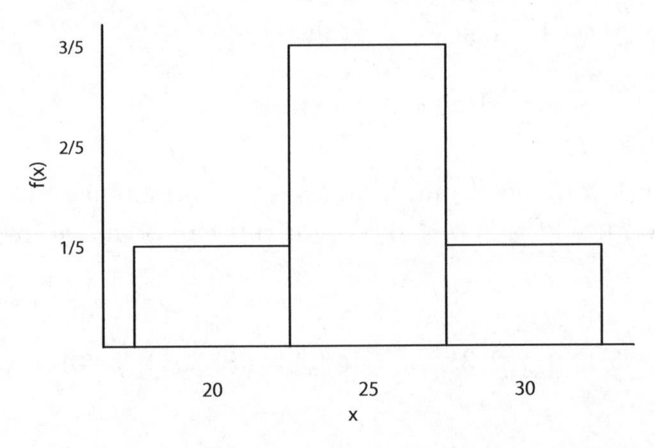

2.19 (a) For $x \ge 0$, $F(x) = \int_0^x \frac{1}{2000}\exp(-t/2000)\, dt = -\exp(-t/2000)\big|_0^x$
$= 1 - \exp(-x/2000)$. So

$$F(x) = \begin{cases} 0, & x < 0, \\ 1 - \exp(-x/2000), & x \ge 0. \end{cases}$$

(b) $P(X > 1000) = 1 - F(1000) = 1 - [1 - \exp(-1000/2000)] = 0.6065$.

(c) $P(X < 2000) = F(2000) = 1 - \exp(-2000/2000) = 0.6321.$

2.21 (a) $f(x) \geq 0$ and $\int_1^\infty 3x^{-4}\,dx = -3\,\frac{x^{-3}}{3}\Big|_1^\infty = 1$. So, this is a valid density function.

(b) For $x \geq 1$, $F(x) = \int_1^x 3t^{-4}\,dt = 1 - x^{-3}$. So,

$$F(x) = \begin{cases} 0, & x < 1, \\ 1 - x^{-3}, & x \geq 1. \end{cases}$$

(c) $P(X > 4) = 1 - F(4) = 4^{-3} = 0.0156.$

2.23 (a) For $y \geq 0$, $F(y) = \frac{1}{4}\int_0^y e^{-t/4}\,dy = 1 - e^{y/4}$. So, $P(Y > 6) = e^{-6/4} = 0.2231$. This probability certainly cannot be considered as "unlikely."

(b) $P(Y \leq 1) = 1 - e^{-1/4} = 0.2212$, which is not so small either.

2.25 (a) Using integral by parts and setting $1 = k\int_0^1 y^4(1-y)^3\,dy$, we obtain $k = 280$.

(b) For $0 \leq y < 1$, $F(y) = 56y^5(1-y)^3 + 28y^6(1-y)^2 + 8y^7(1-y) + y^8$. So, $P(Y \leq 0.5) = 0.3633$.

(c) Using the cdf in (b), $P(Y > 0.8) = 0.0563$.

2.27 (a) $P(X > 8) = 1 - P(X \leq 8) = \sum_{x=0}^{8} e^{-6}\frac{6^x}{x!} = 1 - e^{-6}\left(\frac{6^0}{0!} + \frac{6^1}{1!} + \cdots + \frac{6^8}{8!}\right) = 0.1528.$

(b) $P(X = 2) = e^{-6}\frac{6^2}{2!} = 0.0446.$

2.29 (a) $\sum_{x=0}^{3}\sum_{y=0}^{3} f(x,y) = c\sum_{x=0}^{3}\sum_{y=0}^{3} xy = 36c = 1$. Hence $c = 1/36$.

(b) $\sum_x\sum_y f(x,y) = c\sum_x\sum_y |x-y| = 15c = 1$. Hence $c = 1/15$.

2.31 (a) We can select x oranges from 3, y apples from 2, and $4 - x - y$ bananas from 3 in $\binom{3}{x}\binom{2}{y}\binom{3}{4-x-y}$ ways. A random selection of 4 pieces of fruit can be made in $\binom{8}{4}$ ways. Therefore,

$$f(x,y) = \frac{\binom{3}{x}\binom{2}{y}\binom{3}{4-x-y}}{\binom{8}{4}}, \qquad x = 0,1,2,3; \quad y = 0,1,2; \quad 1 \leq x+y \leq 4.$$

Hence, we have the following joint probability table with the marginal distributions on the last row and last column.

	$f(x,y)$	x = 0	1	2	3	$f_Y(y)$
	0	0	3/70	9/70	3/70	3/14
y	1	2/70	18/70	18/70	2/70	8/14
	2	3/70	9/70	3/70	0	3/14
	$f_X(x)$	1/14	6/14	6/14	1/14	

(b) $P[(X,Y) \in A] = P(X+Y \leq 2) = f(1,0) + f(2,0) + f(0,1) + f(1,1) + f(0,2)$
$= 3/70 + 9/70 + 2/70 + 18/70 + 3/70 = 1/2.$

(c) $P(Y=0|X=2) = \dfrac{P(X=2,Y=0)}{P(X=2)} = \dfrac{9/70}{6/14} = \dfrac{3}{10}.$

(d) We know from (c) that $P(Y=0|X=2) = 3/10$, and we can calculate

$$P(Y=1|X=2) = \frac{18/70}{6/14} = \frac{3}{5}, \text{ and } P(Y=2|X=2) = \frac{3/70}{6/14} = \frac{1}{10}.$$

2.33 (a) $P(X+Y \leq 1/2) = \int_0^{1/2}\int_0^{1/2-y} 24xy\, dx\, dy = 12\int_0^{1/2}\left(\frac{1}{2}-y\right)^2 y\, dy = \frac{1}{16}.$

(b) $g(x) = \int_0^{1-x} 24xy\, dy = 12x(1-x)^2$, for $0 \leq x < 1$.

(c) $f(y|x) = \frac{24xy}{12x(1-x)^2} = \frac{2y}{(1-x)^2}$, for $0 \leq y \leq 1-x$.
Therefore, $P(Y < 1/8 \mid X = 3/4) = 32\int_0^{1/8} y\, dy = 1/4.$

2.35 (a) $P(0 \leq X \leq 1/2,\ 1/4 \leq Y \leq 1/2) = \int_0^{1/2}\int_{1/4}^{1/2} 4xy\, dy\, dx = 3/8\int_0^{1/2} x\, dx = 3/64.$

(b) $P(X < Y) = \int_0^1\int_0^y 4xy\, dx\, dy = 2\int_0^1 y^3\, dy = 1/2.$

2.37 $P(X+Y > 1/2) = 1 - P(X+Y < 1/2) = 1 - \int_0^{1/4}\int_x^{1/2-x} \frac{1}{y}\, dy\, dx$
$= 1 - \int_0^{1/4}\left[\ln\left(\frac{1}{2}-x\right) - \ln x\right]\, dx = 1 + \left[\left(\frac{1}{2}-x\right)\ln\left(\frac{1}{2}-x\right) - x\ln x\right]\Big|_0^{1/4}$
$= 1 + \frac{1}{4}\ln\left(\frac{1}{4}\right) = 0.6534.$
The upper limit of y-integral comes from $x+y < 1/2$ and $0 < x < y$.

2.39 (a)

x	1	2	3
$g(x)$	0.10	0.35	0.55

(b)

y	1	3	5
$h(y)$	0.20	0.50	0.30

(c) $P(Y=3 \mid X=2) = \frac{0.1}{0.05+0.10+0.20} = 0.2857.$

2.41 $g(x) = \frac{1}{8}\int_2^4 (6-x-y)\, dy = \frac{3-x}{4}$, for $0 < x < 2$.
So, $f(y|x) = \frac{f(x,y)}{g(x)} = \frac{6-x-y}{2(3-x)}$, for $2 < y < 4$,
and $P(1 < Y < 3 \mid X = 1) = \frac{1}{4}\int_2^3 (5-y)\, dy = \frac{5}{8}.$

2.43 X and Y are independent since $f(x,y) = g(x)h(y)$ for all (x,y).

2.45 (a) $1 = k\int_0^2\int_0^1\int_0^1 xy^2 z\, dx\, dy\, dz = 2k\int_0^1\int_0^1 y^2 z\, dy\, dz = \frac{2k}{3}\int_0^1 z\, dz = \frac{k}{3}$. So, $k = 3$.

(b) $P\left(X < \frac{1}{4}, Y > \frac{1}{2}, 1 < Z < 2\right) = 3\int_1^2\int_{1/2}^1\int_0^{1/4} xy^2 z\, dx\, dy\, dz = \frac{9}{2}\int_0^{1/4}\int_{1/2}^1 y^2 z\, dy\, dz$
$= \frac{21}{16}\int_1^2 z\, dz = \frac{21}{512}.$

2.47 $g(x) = 4\int_0^1 xy\, dy = 2x$, for $0 < x < 1$; $h(y) = 4\int_0^1 xy\, dx = 2y$, for $0 < y < 1$. Since $f(x,y) = g(x)h(y)$ for all (x,y), X and Y are independent.

2.49 $g(x) = k \int_{30}^{50} (x^2 + y^2)\, dy = k \left(x^2 y + \frac{y^3}{3} \right) \Big|_{30}^{50} = k \left(20x^2 + \frac{98{,}000}{3} \right)$, and
$h(y) = k \left(20y^2 + \frac{98{,}000}{3} \right)$.
Since $f(x, y) \neq g(x)h(y)$, X and Y are not independent.

2.51 $\mu = E(X) = (0)(0.41) + (1)(0.37) + (2)(0.16) + (3)(0.05) + (4)(0.01) = 0.88$.

2.53 $\mu = E(X) = (20)(1/5) + (25)(3/5) + (30)(1/5) = 25$ cents.

2.55 Expected gain $= E(X) = (4000)(0.3) + (-1000)(0.7) = \$500$.

2.57 $E(X) = \frac{4}{\pi} \int_0^1 \frac{x}{1+x^2}\, dx = \frac{\ln 4}{\pi}$.

2.59 $E(X) = \int_0^1 x^2\, dx + \int_1^2 x(2-x)\, dx = 1$. Therefore, the average number of hours per year is $(1)(100) = 100$ hours.

2.61 $E(X) = \frac{1}{\pi a^2} \int_{-a}^{a} \int_{-\sqrt{a^2-y^2}}^{\sqrt{a^2-y^2}} x\, dx\, dy = \frac{1}{\pi a^2} \int_{-a}^{a} \left[\left(\frac{a^2-y^2}{2} \right) - \left(\frac{a^2-y^2}{2} \right) \right] dy = 0$.

2.63 The probability density function is,

x	-3	6	9
$f(x)$	1/6	1/2	1/3
$g(x)$	25	169	361

$\mu_{g(X)} = E[(2X + 1)^2] = (25)(1/6) + (169)(1/2) + (361)(1/3) = 209$.

2.65 Let $Y = 1200X - 50X^2$ be the amount spent.

x	0	1	2	3
$f(x)$	1/10	3/10	2/5	1/5
$y = g(x)$	0	1150	2200	3150

$\mu_Y = E(1200X - 50X^2) = (0)(1/10) + (1150)(3/10) + (2200)(2/5) + (3150)(1/5)$
$= \$1,855$.

2.67 (a) $E[g(X, Y)] = E(XY^2) = \sum_x \sum_y xy^2 f(x, y)$
$= (2)(1)^2(0.10) + (2)(3)^2(0.20) + (2)(5)^2(0.10) + (4)(1)^2(0.15) + (4)(3)^2(0.30)$
$+ (4)(5)^2(0.15) = 35.2$.

(b) $\mu_X = E(X) = (2)(0.40) + (4)(0.60) = 3.20$,
$\mu_Y = E(Y) = (1)(0.25) + (3)(0.50) + (5)(0.25) = 3.00$.

2.69 $E(X) = \frac{1}{2000} \int_0^\infty x \exp(-x/2000)\, dx = 2000 \int_0^\infty y \exp(-y)\, dy = 2000$.

2.71 (a) The density function is shown next

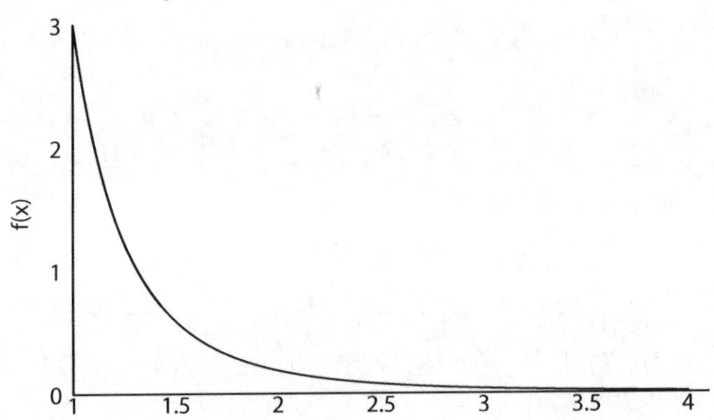

(b) $\mu = E(X) = \int_1^\infty 3x^{-3}\, dx = \frac{3}{2}$.

2.73 (a) $\mu = E(Y) = 5\int_0^1 y(1-y)^4\, dy = -\int_0^1 y\, d(1-y)^5 = \int_0^1 (1-y)^5\, dy = \frac{1}{6}$.

(b) $P(Y > 1/6) = \int_{1/6}^1 5(1-y)^4\, dy = -\left.(1-y)^5\right|_{1/6}^1 = (1-1/6)^5 = 0.4019$.

2.75 (a) A histogram is shown next.

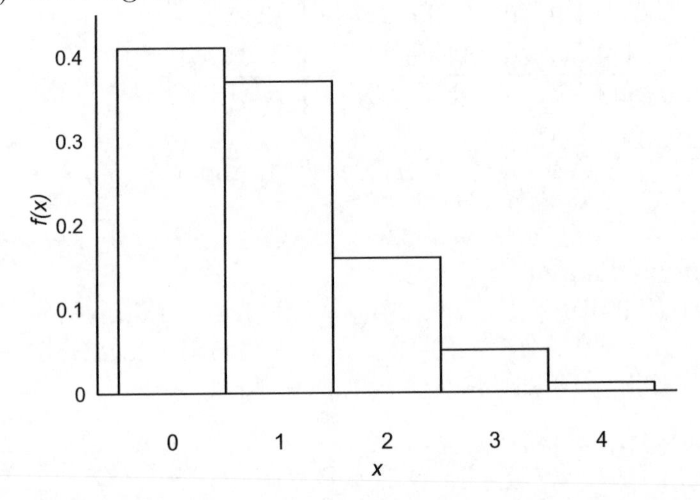

(b) $\mu = (0)(0.41) + (1)(0.37) + (2)(0.16) + (3)(0.05) + (4)(0.01) = 0.88$.

(c) $E(X^2) = (0)^2(0.41) + (1)^2(0.37) + (2)^2(0.16) + (3)^2(0.05) + (4)^2(0.01) = 1.62$.

2.77 $\mu = (2)(0.01) + (3)(0.25) + (4)(0.4) + (5)(0.3) + (6)(0.04) = 4.11$,
$E(X^2) = (2)^2(0.01) + (3)^2(0.25) + (4)^2(0.4) + (5)^2(0.3) + (6)^2(0.04) = 17.63$.
So, $\sigma^2 = 17.63 - 4.11^2 = 0.74$.

2.79 It is know $\mu = 1/3$.
So, $E(X^2) = \int_0^1 2x^2(1-x)\, dx = 1/6$ and $\sigma^2 = 1/6 - (1/3)^2 = 1/18$. So, in the actual profit, the variance is $\frac{1}{18}(5000)^2$.

2.81 It is known $\mu = 1$.
Since $E(X^2) = \int_0^1 x^2 \cdot x\, dx + \int_1^2 x^2(2-x)\, dx = 7/6$, then $\sigma^2 = 7/6 - (1)^2 = 1/6$.

2.83 $\mu_Y = E(3X-2) = \frac{1}{4}\int_0^\infty (3x-2)e^{-x/4}\,dx = 10$. So
$\sigma_Y^2 = E\{[(3X-2)-10]^2\} = \frac{9}{4}\int_0^\infty (x-4)^2 e^{-x/4}\,dx = 144$.

2.85 $g(x) = \frac{2}{3}\int_0^1 (x+2y)\,dy = \frac{2}{3}(x+1)$, for $0 < x < 1$, so $\mu_X = \frac{2}{3}\int_0^1 x(x+1)\,dx = \frac{5}{9}$;
$h(y) = \frac{2}{3}\int_0^1 (x+2y)\,dx = \frac{2}{3}\left(\frac{1}{2}+2y\right)$, so $\mu_Y = \frac{2}{3}\int_0^1 y\left(\frac{1}{2}+2y\right)\,dy = \frac{11}{18}$; and
$E(XY) = \frac{2}{3}\int_0^1\int_0^1 xy(x+2y)\,dy\,dx = \frac{1}{3}$.
So, $\sigma_{XY} = E(XY) - \mu_X\mu_Y = \frac{1}{3} - \left(\frac{5}{9}\right)\left(\frac{11}{18}\right) = -0.0062$.

2.87 $E(X) = (0)(0.41) + (1)(0.37) + (2)(0.16) + (3)(0.05) + (4)(0.01) = 0.88$
and $E(X^2) = (0)^2(0.41) + (1)^2(0.37) + (2)^2(0.16) + (3)^2(0.05) + (4)^2(0.01) = 1.62$.
So, $Var(X) = 1.62 - 0.88^2 = 0.8456$ and $\sigma = \sqrt{0.8456} = 0.9196$.

2.89 The joint and marginal probability mass functions are given in the following table.

		y			
		0	1	2	$f_X(x)$
	0	0	1/35	3/70	1/14
x	1	3/70	9/35	9/70	3/7
	2	9/70	9/35	3/70	3/7
	3	3/70	1/35	0	1/14
	$f_X(x)$	3/14	4/7	3/14	

Hence, $E(X) = \frac{3}{2}$, $E(Y) = 1$, $E(XY) = \frac{9}{7}$, $Var(X) = \frac{15}{28}$, and $Var(Y) = \frac{3}{7}$. Finally, $\rho_{XY} = -\frac{1}{\sqrt{5}}$.

2.91 Let X = number of cartons sold and Y = profit.
We can write $Y = 1.65X + (0.90)(5-X) - (1.20)(5) = 0.75X - 1.50$. Now
$E(X) = (0)(1/15)+(1)(2/15)+(2)(2/15)+(3)(3/15)+(4)(4/15)+(5)(3/15) = 46/15$,
and $E(Y) = (0.75)E(X) - 1.50 = (0.75)(46/15) - 1.50 = \$0.80$.

2.93 The equations $E[(X-1)^2] = 10$ and $E[(X-2)^2] = 6$ may be written in the form:

$$E(X^2) - 2E(X) = 9, \qquad E(X^2) - 4E(X) = 2.$$

Solving these two equations simultaneously we obtain

$$E(X) = 7/2, \quad \text{and} \quad E(X^2) = 16.$$

Hence $\mu = 7/2$ and $\sigma^2 = 16 - (7/2)^2 = 15/4$.

2.95 $E(2XY^2 - X^2Y) = 2E(XY^2) - E(X^2Y)$. Now,
$E(XY^2) = \sum_{x=0}^{2}\sum_{y=0}^{2} xy^2 f(x,y) = (1)(1)^2(3/14) = 3/14$, and
$E(X^2Y) = \sum_{x=0}^{2}\sum_{y=0}^{2} x^2 y f(x,y) = (1)^2(1)(3/14) = 3/14$.
Therefore, $E(2XY^2 - X^2Y) = (2)(3/14) - (3/14) = 3/14$.

2.97 $\sigma_Z^2 = \sigma_{-2X+4Y-3}^2 = 4\sigma_X^2 + 16\sigma_Y^2 - 16\sigma_{XY} = (4)(5) + (16)(3) - (16)(1) = 52.$

2.99 For $0 < a < 1$, since $g(a) = \sum_{x=0}^{\infty} a^x = \frac{1}{1-a}$, $g'(a) = \sum_{x=1}^{\infty} x a^{x-1} = \frac{1}{(1-a)^2}$ and $g''(a) = \sum_{x=2}^{\infty} x(x-1)a^{x-2} = \frac{2}{(1-a)^3}$. These results will be used to evaluate the sums below.

(a) The marginal distributions $g(x)$ and $h(y)$ are given in Review Exercise 2.118.
$E(X) = (3/4)\sum_{x=1}^{\infty} x(1/4)^x = (3/4)(1/4)\sum_{x=1}^{\infty} x(1/4)^{x-1} = (3/16)[1/(1-1/4)^2]$
$= 1/3$, and $E(Y) = E(X) = 1/3$.
$E(X^2) - E(X) = E[X(X-1)] = (3/4)\sum_{x=2}^{\infty} x(x-1)(1/4)^x$
$= (3/4)(1/4)^2 \sum_{x=2}^{\infty} x(x-1)(1/4)^{x-2} = (3/4^3)[2/(1-1/4)^3] = 2/9.$
So, $Var(X) = E(X^2) - [E(X)]^2 = [E(X^2) - E(X)] + E(X) - [E(X)]^2$
$2/9 + 1/3 - (1/3)^2 = 4/9$, and $Var(Y) = 4/9$.

(b) $E(Z) = E(X) + E(Y) = (1/3) + (1/3) = 2/3$, and
$Var(Z) = Var(X+Y) = Var(X) + Var(Y) = (4/9) + (4/9) = 8/9$, since X and Y are independent (from Exercise 3.79).

2.101 (a) $E(Y) = \int_0^\infty y e^{-y/4}\, dy = 4.$

(b) $E(Y^2) = \int_0^\infty y^2 e^{-y/4}\, dy = 32$ and $Var(Y) = 32 - 4^2 = 16.$

Chapter 3

Some Probability Distributions

3.1 By definition, we have

$$\mu = E(X) = \sum_{i=1}^{k} x_i \frac{1}{k} = \frac{1}{k}\sum_{i=1}^{k} x_i,$$

and

$$\sigma^2 = \frac{1}{k}\sum_{i=1}^{k}(x_i - \mu)^2.$$

3.3 This is a uniform distribution: $f(x) = \frac{1}{10}$, for $x = 1, 2, \dots, 10$, and $f(x) = 0$ elsewhere. Therefore $P(X < 4) = \sum_{x=1}^{3} f(x) = \frac{3}{10}$.

3.5 $p = 0.7$.

(a) For $n = 10$, $P(X < 5) = P(X \leq 4) = 0.0474$.

(b) For $n = 20$, $P(X < 10) = P(X \leq 9) = 0.0171$.

3.7 For $n = 15$ and $p = 0.25$, we have

(a) $P(3 \leq X \leq 6) = P(X \leq 6) - P(X \leq 2) = 0.9434 - 0.2361 = 0.7073$.

(b) $P(X < 4) = P(X \leq 3) = 0.4613$.

(c) $P(X > 5) = 1 - P(X \leq 5) = 1 - 0.8516 = 0.1484$.

3.9 From Table A.1 with $n = 7$ and $p = 0.9$, we have
$P(X = 5) = P(X \leq 5) - P(X \leq 4) = 0.1497 - 0.0257 = 0.1240$.

3.11 $p = 0.4$ and $n = 5$.

(a) $P(X = 0) = 0.0778$.

(b) $P(X < 2) = P(X \le 1) = 0.3370$.

(c) $P(X > 3) = 1 - P(X \le 3) = 1 - 0.9130 = 0.0870$.

3.13 Let X_1 = number of times encountered green light with $P(\text{Green}) = 0.35$, X_2 = number of times encountered yellow light with $P(\text{Yellow}) = 0.05$, and X_3 = number of times encountered red light with $P(\text{Red}) = 0.60$. Then

$$f(x_1, x_2, x_3) = \binom{n}{x_1, x_2, x_3}(0.35)^{x_1}(0.05)^{x_2}(0.60)^{x_3}.$$

3.15 (a) $\binom{10}{2,5,3}(0.225)^2(0.544)^5(0.231)^3 = 0.0749$.

(b) $\binom{10}{10}(0.544)^{10}(0.456)^0 = 0.0023$.

(c) $\binom{10}{0}(0.225)^0(0.775)^{10} = 0.0782$.

3.17 $n = 20$ and the probability of a defective is $p = 0.10$. So, $P(X \le 3) = 0.8670$.

3.19 $n = 20$ and $p = 0.90$;

(a) $P(X = 18) = P(X \le 18) - P(X \le 17) = 0.6083 - 0.3231 = 0.2852$.

(b) $P(X \ge 15) = 1 - P(X \le 14) = 1 - 0.0113 = 0.9887$.

(c) $P(X \le 18) = 0.6083$.

3.21 $P(X \ge 1) = 1 - P(X = 0) = 1 - h(0; 15, 3, 6) = 1 - \frac{\binom{6}{0}\binom{9}{3}}{\binom{15}{3}} = \frac{53}{65}$.

3.23 $P(X \le 2) = \sum_{x=0}^{2} h(x; 50, 5, 10) = 0.9517$.

3.25 Using the binomial approximation of the hypergeometric distribution with $p = 30/150 = 0.2$, the probability is $1 - \sum_{x=0}^{2} b(x; 10, 0.2) = 0.3222$.

3.27 (a) $P(X = 0) = b(0; 3, 3/25) = 0.6815$.

(b) $P(1 \le X \le 3) = \sum_{x=1}^{3} b(x; 3, 1/25) = 0.1153$.

3.29 $h(5; 25, 15, 10) = \frac{\binom{10}{5}\binom{15}{10}}{\binom{25}{15}} = 0.2315$.

3.31 (a) $\frac{\binom{3}{0}\binom{17}{5}}{\binom{20}{5}} = 0.3991$.

(b) $\frac{\binom{3}{2}\binom{17}{3}}{\binom{20}{5}} = 0.1316$.

3.33 $N = 10000$, $n = 30$ and $k = 300$. Using binomial approximation to the hypergeometric distribution with $p = 300/10000 = 0.03$, the probability of $\{X \geq 1\}$ can be determined by

$$1 - b(0; 30, 0.03) = 1 - (0.97)^{30} = 0.599.$$

3.35 The probability that all coins turn up the same is $1/4$. Using the geometric distribution with $p = 3/4$ and $q = 1/4$, we have

$$P(X < 4) = \sum_{x=1}^{3} g(x; 3/4) = \sum_{x=1}^{3} (3/4)(1/4)^{x-1} = \frac{63}{64}.$$

3.37 (a) $P(X > 5) = \sum_{x=6}^{\infty} p(x; 5) = 1 - \sum_{x=0}^{5} p(x; 5) = 0.3840.$

(b) $P(X = 0) = p(0; 5) = 0.0067.$

3.39 Using the geometric distribution

(a) $P(X = 3) = g(3; 0.7) = (0.7)(0.3)^2 = 0.0630.$

(b) $P(X < 4) = \sum_{x=1}^{3} g(x; 0.7) = \sum_{x=1}^{3} (0.7)(0.3)^{x-1} = 0.9730.$

3.41 (a) $P(X \geq 4) = 1 - P(X \leq 3) = 0.1429.$

(b) $P(X = 0) = p(0; 2) = 0.1353.$

3.43 $\mu = np = (10000)(0.001) = 10$, so

$$P(6 \leq X \leq 8) = P(X \leq 8) - P(X \leq 5) \approx \sum_{x=0}^{8} p(x; 10) - \sum_{x=0}^{5} p(x; 10) = 0.2657.$$

3.45 $\mu = 6$ and $\sigma^2 = 6$.

3.47 (a) $P(X \leq 3 | \lambda t = 5) = 0.2650.$

(b) $P(X > 1 | \lambda t = 5) = 1 - 0.0404 = 0.9596.$

3.49 (a) $P(X > 10 | \lambda t = 14) = 1 - 0.1757 = 0.8243.$

(b) $\lambda t = 14$.

3.51 $\mu = (4000)(0.001) = 4$.

3.53 $\mu = \lambda t = (1.5)(5) = 7.5$ and $P(X = 0 | \lambda t = 7.5) = e^{-7.5} = 5.53 \times 10^{-4}$.

3.55 (a) $P(X > 10 | \lambda t = 5) = 1 - P(X \leq 10 | \lambda t = 5) = 1 - 0.9863 = 0.0137.$

(b) $\mu = \lambda t = (5)(3) = 15$, so $P(X > 20|\lambda t = 15) = 1 - P(X \leq 20|\lambda = 15) = 1 - 0.9170 = 0.0830$.

3.57 $f(x) = \frac{1}{B-A}$ for $A \leq x \leq B$.

(a) $\mu = \int_A^B \frac{x}{B-A}\, dx = \frac{B^2-A^2}{2(B-A)} = \frac{A+B}{2}$.

(b) $E(X^2) = \int_A^B \frac{x^2}{B-A}\, dx = \frac{B^3-A^3}{3(B-A)}$.
So, $\sigma^2 = \frac{B^3-A^3}{3(B-A)} - \left(\frac{A+B}{2}\right)^2 = \frac{4(B^2+AB+A^2)-3(B^2+2AB+A^2)}{12} = \frac{B^2-2AB+A^2}{12} = \frac{(B-A)^2}{12}$.

3.59 $A = 7$ and $B = 10$.

(a) $P(X \leq 8.8) = \frac{8.8-7}{3} = 0.60$.

(b) $P(7.4 < X < 9.5) = \frac{9.5-7.4}{3} = 0.70$.

(c) $P(X \geq 8.5) = \frac{10-8.5}{3} = 0.50$.

3.61 (a) Area=0.0823.

(b) Area=$1 - 0.9750 = 0.0250$.

(c) Area=$0.2578 - 0.0154 = 0.2424$.

(d) Area=0.9236.

(e) Area=$1 - 0.1867 = 0.8133$.

(f) Area=$0.9591 - 0.3156 = 0.6435$.

3.63 (a) $z = (15 - 18)/2.5 = -1.2$; $P(X < 15) = P(Z < -1.2) = 0.1151$.

(b) $z = -0.76$, $k = (2.5)(-0.76) + 18 = 16.1$.

(c) $z = 0.91$, $k = (2.5)(0.91) + 18 = 20.275$.

(d) $z_1 = (17 - 18)/2.5 = -0.4$, $z_2 = (21 - 18)/2.5 = 1.2$;
$P(17 < X < 21) = P(-0.4 < Z < 1.2) = 0.8849 - 0.3446 = 0.5403$.

3.65 (a) $z = (224 - 200)/15 = 1.6$. Fraction of the cups containing more than 224 millimeters is $P(Z > 1.6) = 0.0548$.

(b) $z_1 = (191 - 200)/15 = -0.6$, $Z_2 = (209 - 200)/15 = 0.6$;
$P(191 < X < 209) = P(-0.6 < Z < 0.6) = 0.7257 - 0.2743 = 0.4514$.

(c) $z = (230 - 200)/15 = 2.0$; $P(X > 230) = P(Z > 2.0) = 0.0228$. Therefore, $(1000)(0.0228) = 22.8$ or approximately 23 cups will overflow.

(d) $z = -0.67$, $x = (15)(-0.67) + 200 = 189.95$ millimeters.

3.67 (a) $z = (32 - 40)/6.3 = -1.27$; $P(X > 32) = P(Z > -1.27) = 1 - 0.1020 = 0.8980$.

(b) $z = (28 - 40)/6.3 = -1.90$, $P(X < 28) = P(Z < -1.90) = 0.0287$.

(c) $z_1 = (37 - 40)/6.3 = -0.48$, $z_2 = (49 - 40)/6.3 = 1.43$;
So, $P(37 < X < 49) = P(-0.48 < Z < 1.43) = 0.9236 - 0.3156 = 0.6080$.

3.69 (a) $z = (30 - 24)/3.8 = 1.58$; $P(X > 30) = P(Z > 1.58) = 0.0571$.

(b) $z = (15 - 24)/3.8 = -2.37$; $P(X > 15) = P(Z > -2.37) = 0.9911$. He is late 99.11% of the time.

(c) $z = (25 - 24)/3.8 = 0.26$; $P(X > 25) = P(Z > 0.26) = 0.3974$.

(d) $z = 1.04$, $x = (3.8)(1.04) + 24 = 27.952$ minutes.

(e) Using the binomial distribution with $p = 0.0571$, we get $b(2; 3, 0.0571) = \binom{3}{2}(0.0571)^2(0.9429) = 0.0092$.

3.71 $z = -1.88$, $x = (2)(-1.88) + 10 = 6.24$ years.

3.73 (a) $z = (10,175 - 10,000)/100 = 1.75$. Proportion of components exceeding 10.150 kilograms in tensile strength$= P(X > 10,175) = P(Z > 1.75) = 0.0401$.

(b) $z_1 = (9,775 - 10,000)/100 = -2.25$ and $z_2 = (10,225 - 10,000)/100 = 2.25$. Proportion of components scrapped$= P(X < 9,775) + P(X > 10,225) = P(Z < -2.25) + P(Z > 2.25) = 2P(Z < -2.25) = 0.0244$.

3.75 $z = (94.5 - 115)/12 = -1.71$; $P(X < 94.5) = P(Z < -1.71) = 0.0436$. Therefore, $(0.0436)(600) = 26$ students will be rejected.

3.77 $n = 100$.

(a) $p = 0.01$ with $\mu = (100)(0.01) = 1$ and $\sigma = \sqrt{(100)(0.01)(0.99)} = 0.995$. So, $z = (0.5 - 1)/0.995 = -0.503$. $P(X \le 0) \approx P(Z \le -0.503) = 0.3085$.

(b) $p = 0.05$ with $\mu = (100)(0.05) = 5$ and $\sigma = \sqrt{(100)(0.05)(0.95)} = 2.1794$. So, $z = (0.5 - 5)/2.1794 = -2.06$. $P(X \le 0) \approx P(X \le -2.06) = 0.0197$.

3.79 $\mu = (100)(0.9) = 90$ and $\sigma = \sqrt{(100)(0.9)(0.1)} = 3$.

(a) $z_1 = (83.5 - 90)/3 = -2.17$ and $z_2 = (95.5 - 90)/3 = 1.83$. $P(83.5 < X < 95.5) = P(-2.17 < Z < 1.83) = 0.9664 - 0.0150 = 0.9514$.

(b) $z = (85.5 - 90)/3 = -1.50$; $P(X < 85.5) = P(Z < -1.50) = 0.0668$.

3.81 (a) $p = 0.05$, $n = 100$ with $\mu = 5$ and $\sigma = \sqrt{(100)(0.05)(0.95)} = 2.1794$. So, $z = (2.5 - 5)/2.1794 = -1.147$; $P(X \ge 2) \approx P(Z \ge -1.147) = 0.8749$.

(b) $z = (10.5 - 5)/2.1794 = 2.524$; $P(X \ge 10) \approx P(Z > 2.52) = 0.0059$.

3.83 $\mu = (400)(1/10) = 40$ and $\sigma = \sqrt{(400)(1/10)(9/10)} = 6$.

(a) $z = (31.5 - 40)/6 = -1.42$; $P(X < 31.5) = P(Z < -1.42) = 0.0778$.

(b) $z = (49.5 - 40)/6 = 1.58$; $P(X > 49.5) = P(Z > 1.58) = 1 - 0.9429 = 0.0571$.

(c) $z_1 = (34.5 - 40)/6 = -0.92$ and $z_2 = (46.5 - 40)/6 = 1.08$; $P(34.5 < X < 46.5) = P(-0.92 < Z < 1.08) = 0.8599 - 0.1788 = 0.6811$.

3.85 (a) $P(X \ge 230) = P\left(Z > \frac{230-170}{30}\right) = 0.0228$.

(b) Denote by Y the number of students whose serum cholesterol level exceed 230 among the 300. Then $Y \sim b(y; 300, 0.0228$ with

$$\mu = (300)(0.0228) = 6.84, \text{ and } \sigma = \sqrt{(300)(0.0228)(1 - 0.0228)} = 2.5854.$$

So, $z = \frac{8-0.5-6.84}{2.5854} = 0.26$ and
$P(X \geq 8) \approx P(Z > 0.26) = 0.3974$.

3.87 (a) Denote by X the number of failures among the 20. $X \sim b(x; 20, 0.01)$ and $P(X > 1) = 1 - b(0; 20, 0.01) - b(1; 20, 0.01) = 1 - \binom{20}{0}(0.01)^0(0.99)^{20} - \binom{20}{1}(0.01)(0.99)^{19} = 0.01686$.

(b) $n = 500$ and $p = 0.01$ with $\mu = (500)(0.01) = 5$ and $\sigma = \sqrt{(500)(0.01)(0.99)} = 2.2249$. So, $P(\text{more than 8 failures}) \approx P(Z > (8.5-5)/2.2249) = P(Z > 1.57) = 1 - 0.9418 = 0.0582$.

3.89 $P(1.8 < X < 2.4) = \int_{1.8}^{2.4} xe^{-x}\, dx = [-xe^{-x} - e^{-x}]|_{1.8}^{2.4} = 2.8e^{-1.8} - 3.4e^{-2.4} = 0.1545$.

3.91 $\mu = \alpha\beta = (2)(3) = 6$ million liters; $\sigma^2 = \alpha\beta^2 = (2)(9) = 18$.

3.93 $P(X < 3) = \frac{1}{4}\int_0^3 e^{-x/4}\, dx = -e^{-x/4}\big|_0^3 = 1 - e^{-3/4} = 0.5276$.
Let Y be the number of days a person is served in less than 3 minutes. Then
$P(Y \geq 4) = \sum_{x=4}^{6} b(y; 6, 1 - e^{-3/4}) = \binom{6}{4}(0.5276)^4(0.4724)^2 + \binom{6}{5}(0.5276)^5(0.4724)$
$+ \binom{6}{6}(0.5276)^6 = 0.3968$.

3.95 $\alpha = 5$; $\beta = 10$;

(a) $\alpha\beta = 50$.

(b) $\sigma^2 = \alpha\beta^2 = 500$; so $\sigma = \sqrt{500} = 22.36$.

(c) $P(X > 30) = \frac{1}{\beta^\alpha\Gamma(\alpha)}\int_{30}^{\infty} x^{\alpha-1}e^{-x/\beta}\, dx$. Using the incomplete gamma with $y = x/\beta$, then

$$1 - P(X \leq 30) = 1 - P(Y \leq 3) = 1 - \int_0^3 \frac{y^4e^{-y}}{\Gamma(5)}\, dy = 1 - 0.185 = 0.815.$$

3.97 $\mu = 3$ seconds with $f(x) = \frac{1}{3}e^{-x/3}$ for $x > 0$.

(a) $P(X > 5) = \int_5^{\infty} \frac{1}{3}e^{-x/3}\, dx = \frac{1}{3}\left[-3e^{-x/3}\right]\big|_5^{\infty} = e^{-5/3} = 0.1889$.

(b) $P(X > 10) = e^{-10/3} = 0.0357$.

3.99 Let T be the time between two consecutive arrivals

(a) $P(T > 1) = P(\text{no arrivals in 1 minute}) = P(X = 0) = e^{-5} = 0.0067$.

(b) $\mu = \beta = 1/5 = 0.2$.

Chapter 4

Sampling Distributions and Data Descriptions

4.1 (a) Responses of all people in Richmond who have telephones.

(b) Outcomes for a large or infinite number of tosses of a coin.

(c) Length of life of such tennis shoes when worn on the professional tour.

(d) All possible time intervals for this lawyer to drive from her home to her office.

4.3 (a) $\bar{x} = 53.75$.

(b) Modes are 75 and 100.

4.5 (a) Range $= 15 - 5 = 10$.

(b) $s^2 = \frac{n\sum_{i=1}^{n} x_i^2 - (\sum_{i=1}^{n} x_i)^2}{n(n-1)} = \frac{(10)(838)-86^2}{(10)(9)} = 10.933$. Taking the square root, we have $s = 3.307$.

4.7 (a) $s^2 = \frac{1}{n-1}\sum_{x=1}^{n}(x_i - \bar{x})^2 = \frac{1}{11}[(48-35.7)^2 + (47-35.7)^2 + \cdots + (26-35.7)^2] = 61.2$.

(b) $s^2 = \frac{n\sum_{i=1}^{n} x_i^2 - (\sum_{i=1}^{n} x_i)^2}{n(n-1)} = \frac{(12)(15398)-428^2}{(12)(11)} = 61.2$.

4.9 $s^2 = \frac{n\sum_{i=1}^{n} x_i^2 - (\sum_{i=1}^{n} x_i)^2}{n(n-1)} = \frac{(20)(148.55)-53.3^2}{(20)(19)} = 0.342$ and hence $s = 0.585$.

4.11 $s^2 = \frac{n\sum_{i=1}^{n} x_i^2 - (\sum_{i=1}^{n} x_i)^2}{n(n-1)} = \frac{(6)(207)-33^2}{(6)(5)} = 5.1$.

(a) Multiplying each observation by 3 gives $s^2 = (9)(5.1) = 45.9$.

(b) Adding 5 to each observation does not change the variance. Hence $s^2 = 5.1$.

4.13 $z_1 = -1.9$, $z_2 = -0.4$. Hence,

$$P(\mu_{\bar{X}} - 1.9\sigma_{\bar{X}} < \bar{X} < \mu_{\bar{X}} - 0.4\sigma_{\bar{X}}) = P(-1.9 < Z < -0.4) = 0.3446 - 0.0287 = 0.3159.$$

4.15 $\mu_{\bar{X}} = \mu = 240$, $\sigma_{\bar{X}} = 15/\sqrt{40} = 2.372$. Therefore, $\mu_{\bar{X}} \pm 2\sigma_{\bar{X}} = 240 \pm (2)(2.372)$ or from 235.256 to 244.744, which indicates that a value of $x = 236$ milliliters is reasonable and hence the machine need not be adjusted.

4.17 (a) $\mu = \sum xf(x) = (4)(0.2) + (5)(0.4) + (6)(0.3) + (7)(0.1) = 5.3$, and $\sigma^2 = \sum(x-\mu)^2 f(x) = (4-5.3)^2(0.2) + (5-5.3)^2(0.4) + (6-5.3)^2(0.3) + (7-5.3)^2(0.1) = 0.81$.

(b) With $n = 36$, $\mu_{\bar{X}} = \mu = 5.3$ and $\sigma^2_{\bar{X}} = \sigma^2/n = 0.81/36 = 0.0225$.

(c) $n = 36$, $\mu_{\bar{X}} = 5.3$, $\sigma_{\bar{X}} = 0.9/6 = 0.15$, and $z = (5.5 - 5.3)/0.15 = 1.33$. So,

$$P(\bar{X} < 5.5) = P(Z < 1.33) = 0.9082.$$

4.19 (a) $P(6.4 < \bar{X} < 7.2) = P(-1.8 < Z < 0.6) = 0.6898$.

(b) $z = 1.04$, $\bar{x} = z(\sigma/\sqrt{n}) + \mu = (1.04)(1/3) + 7 = 7.35$.

4.21 $n = 50$, $\bar{x} = 0.23$ and $\sigma = 0.1$. Now, $z = (0.23 - 0.2)/(0.1/\sqrt{50}) = 2.12$; so

$$P(\bar{X} \geq 0.23) = P(Z \geq 2.12) = 0.0170.$$

Hence the probability of having such observations, given the mean $\mu = 0.20$, is small. Therefore, the mean amount to be 0.20 is not likely to be true.

4.23 (a) If the two population mean drying times are truly equal, the probability that the difference of the two sample means is 1.0 is 0.0013, which is very small. This means that the assumption of the equality of the population means are not reasonable.

(b) If the experiment was run 10,000 times, there would be $(10000)(0.0013) = 13$ experiments where $\bar{X}_A - \bar{X}_B$ would be at least 1.0.

4.25 (a) When the population equals the limit, the probability of a sample mean exceeding the limit would be 1/2 due the symmetry of the approximated normal distribution.

(b) $P(\bar{X} \geq 7960 \mid \mu = 7950) = P(Z \geq (7960 - 7950)/(100/\sqrt{25})) = P(Z \geq 0.5) = 0.3085$. No, this is not very strong evidence that the population mean of the process exceeds the government limit.

4.27 Since the probability that $\bar{X} \leq 775$ is 0.0062, given that $\mu = 800$ is true, it suggests that this event is very rare and it is very likely that the claim of $\mu = 800$ is not true. On the other hand, if μ is truly, say, 760, the probability

$$P(\bar{X} \leq 775 \mid \mu = 760) = P(Z \leq (775 - 760)/(40/\sqrt{16})) = P(Z \leq 1.5) = 0.9332,$$

which is very high.

4.29 (a) 27.488.
(b) 18.475.
(c) 36.415.

4.31 (a) $\chi^2_\alpha = \chi^2_{0.99} = 0.297$.
(b) $\chi^2_\alpha = \chi^2_{0.025} = 32.852$.
(c) $\chi^2_{0.05} = 37.652$. Therefore, $\alpha = 0.05 - 0.045 = 0.005$. Hence, $\chi^2_\alpha = \chi^2_{0.005} = 46.928$.

4.33 (a) $P(S^2 > 9.1) = P\left(\frac{(n-1)S^2}{\sigma^2} > \frac{(24)(9.1)}{6}\right) = P(\chi^2 > 36.4) = 0.05$.
(b) $P(3.462 < S^2 < 10.745) = P\left(\frac{(24)(3.462)}{6} < \frac{(n-1)S^2}{\sigma^2} < \frac{(24)(10.745)}{6}\right)$
$= P(13.848 < \chi^2 < 42.980) = 0.95 - 0.01 = 0.94$.

4.35 (a) $P(T < 2.365) = 1 - 0.025 = 0.975$.
(b) $P(T > 1.318) = 0.10$.
(c) $P(T < 2.179) = 1 - 0.025 = 0.975$, $P(T < -1.356) = P(T > 1.356) = 0.10$. Therefore, $P(-1.356 < T < 2.179) = 0.975 - 0.010 = 0.875$.
(d) $P(T > -2.567) = 1 - P(T > 2.567) = 1 - 0.01 = 0.99$.

4.37 (a) From Table A.4 we note that 2.069 corresponds to $t_{0.025}$ when $v = 23$. Therefore, $-t_{0.025} = -2.069$ which means that the total area under the curve to the left of $t = k$ is $0.025 + 0.965 = 0.990$. Hence, $k = t_{0.01} = 2.500$.
(b) From Table A.4 we note that 2.807 corresponds to $t_{0.005}$ when $v = 23$. Therefore the total area under the curve to the right of $t = k$ is $0.095 + 0.005 = 0.10$. Hence, $k = t_{0.10} = 1.319$.
(c) $t_{0.05} = 1.714$ for 23 degrees of freedom.

4.39 From Table A.4 we find $t_{0.025} = 2.131$ for $v = 15$ degrees of freedom. Since the value

$$t = \frac{27.5 - 30}{5/4} = -2.00$$

falls between -2.131 and 2.131, the claim is valid.

4.41 (a) 2.71.
(b) 3.51.
(c) 2.92.
(d) $1/2.11 = 0.47$.
(e) $1/2.90 = 0.34$.

4.43 $s_1^2 = 15750$ and $s_2^2 = 10920$ which gives $f = 1.44$. Since, from Table A.6, $f_{0.05}(4, 5) = 5.19$, the probability of $F > 1.44$ is much bigger than 0.05, which means that the conjecture that the two population variances are equal cannot be rejected. The actual probability of $F > 1.44$ is 0.3442 and $P(F < 1/1.44) + P(F > 1.44) = 0.7170$.

Chapter 5

One- and Two-Sample Estimation Problems

5.1 $n = [(2.575)(5.8)/2]^2 = 56$ when rounded up.

5.3 $n = 75, \bar{x} = 0.310, \sigma = 0.0015$, and $z_{0.025} = 1.96$. A 95% confidence interval for the population mean is

$$0.310 - (1.96)(0.0015/\sqrt{75}) < \mu < 0.310 + (1.96)(0.0015/\sqrt{75}),$$

or $0.3097 < \mu < 0.3103$.

5.5 $n = 100, \bar{x} = 23,500, \sigma = 3900$, and $z_{0.005} = 2.575$.

(a) A 99% confidence interval for the population mean is
$23,500 - (2.575)(3900/10) < \mu < 23,500 + (2.575)(3900/10)$, or
$22,496 < \mu < 24,504$.

(b) $e < (2.575)(3900/10) = 1004$.

5.7 $n = [(1.96)(0.0015)/0.0005]^2 = 35$ when rounded up.

5.9 $n = 20, \bar{x} = 11.3, s = 2.45$, and $t_{0.025} = 2.093$ with 19 degrees of freedom. A 95% confidence interval for the population mean is

$$11.3 - (2.093)(2.45/\sqrt{20}) < \mu < 11.3 + (2.093)(2.45/\sqrt{20}),$$

or $10.15 < \mu < 12.45$.

5.11 $n = 12, \bar{x} = 48.50, s = 1.5$, and $t_{0.05} = 1.796$ with 11 degrees of freedom. A 90% confidence interval for the population mean is

$$48.50 - (1.796)(1.5/\sqrt{12}) < \mu < 48.50 + (1.796)(1.5/\sqrt{12}),$$

or $47.722 < \mu < 49.278$.

5.13 $n = 100, \bar{x} = 23{,}500, s = 3,900, 1 - \alpha = 0.99$, and $t_{0.005} \approx 2.66$ with 60 degrees of freedom (use table from the book) or $t_{0.005} = 2.626$ if 100 degrees of freedom is used. The prediction interval of next automobile will be driven in Virginia (using 2.66) is $23{,}500 \pm (2.66)(3{,}900)\sqrt{1 + 1/100}$ which yields $13{,}075 < \mu < 33{,}925$ kilometers.

5.15 $n = 25, \bar{x} = 325.05, s = 0.5, \gamma = 5\%$, and $1 - \alpha = 90\%$, with $k = 2.208$. So, $325.05 \pm (2.208)(0.5)$ yields $(323.946, 326.151)$. Thus, we are 95% confident that this tolerance interval will contain 90% of the aspirin contents for this brand of buffered aspirin.

5.17 $n = 15, \bar{x} = 3.84$, and $s = 3.07$. To calculate an upper 95% prediction limit, we obtain $t_{0.05} = 1.761$ with 14 degrees of freedom. So, the upper limit is $3.84 + (1.761)(3.07)\sqrt{1 + 1/15} = 3.84 + 5.58 = 9.42$. This means that a new observation will have a chance of 95% to fall into the interval $(-\infty, 9.42)$. To obtain an upper 95% tolerance limit, using $1 - \alpha = 0.95$ and $\gamma = 0.05$, with $k = 2.566$, we get $3.84 + (2.566)(3.07) = 11.72$. Hence, we are 95% confident that $(-\infty, 11.72)$ will contain 95% of the orthophosphorous measurements in the river.

5.19 (a) $n = 16$, $\bar{x} = 1.0025$, $s = 0.0202$, $1 - \alpha = 0.99$, $t_{0.005} = 2.947$ with 15 degrees of freedom. A 99% confidence interval for the mean diameter is $1.0025 \pm (2.947)(0.0202)/\sqrt{16}$ which yields $(0.9876, 1.0174)$ centimeters.

(b) A 99% prediction interval for the diameter of a new metal piece is

$$1.0025 \pm (2.947)(0.0202)\sqrt{1 + 1/16}$$

which yields $(0.9411, 1.0639)$ centimeters.

(c) For $n = 16$, $1 - \gamma = 0.99$ and $1 - \alpha = 0.95$, we find $k = 3.421$. Hence, the tolerance limits are $1.0025 \pm (3.421)(0.0202)$ which give $(0.9334, 1.0716)$.

5.21 Since the manufacturer would be more interested in the mean tensile strength for future products, it is conceivable that prediction interval and tolerance interval may be more interesting than just a confidence interval.

5.23 In Exercise 9.14, a 95% prediction interval for a new observation is calculated as $(1.6358, 5.9377)$. Since 6.9 is in the outside range of the prediction interval, this new observation is likely to be an outlier.

5.25 $n_1 = 100, n_2 = 200, \bar{x}_1 = 12.2, \bar{x}_2 = 9.1, s_1 = 1.1$, and $s_2 = 0.9$. It is known that $z_{0.01} = 2.327$. So

$$(12.2 - 9.1) \pm 2.327\sqrt{1.1^2/100 + 0.9^2/200} = 3.1 \pm 0.30,$$

or $2.80 < \mu_1 - \mu_2 < 3.40$. The treatment appears to reduce the mean amount of metal removed.

5.27 $n_1 = 12, n_2 = 10, \bar{x}_1 = 85, \bar{x}_2 = 81, s_1 = 4, s_2 = 5$, and $s_p = 4.478$ with $t_{0.05} = 1.725$ with 20 degrees of freedom. So

$$(85 - 81) \pm (1.725)(4.478)\sqrt{1/12 + 1/10} = 4 \pm 3.31,$$

which yields $0.69 < \mu_1 - \mu_2 < 7.31$.

5.29 $n_1 = 14, n_2 = 16, \bar{x}_1 = 17, \bar{x}_2 = 19, s_1^2 = 1.5, s_2^2 = 1.8$, and $s_p = 1.289$ with $t_{0.005} = 2.763$ with 28 degrees of freedom. So,

$$(19 - 17) \pm (2.763)(1.289)\sqrt{1/16 + 1/14} = 2 \pm 1.30,$$

which yields $0.70 < \mu_1 - \mu_2 < 3.30$.

5.31 $n_A = n_B = 12, \bar{x}_A = 36,300, \bar{x}_B = 38,100, s_A = 5,000, s_B = 6,100$, and

$$v = \frac{5000^2/12 + 6100^2/12}{\frac{(5000^2/12)^2}{11} + \frac{(6100^2/12)^2}{11}} = 21,$$

with $t_{0.025} = 2.080$ with 21 degrees of freedom. So,

$$(36,300 - 38,100) \pm (2.080)\sqrt{\frac{5000^2}{12} + \frac{6100^2}{12}} = -1,800 \pm 4,736,$$

which yields $-6,536 < \mu_A - \mu_B < 2,936$.

5.33 $n = 9, \bar{d} = 2.778, s_d = 4.5765$, with $t_{0.025} = 2.306$ with 8 degrees of freedom. So,

$$2.778 \pm (2.306)\frac{4.5765}{\sqrt{9}} = 2.778 \pm 3.518,$$

which yields $-0.74 < \mu_D < 6.30$.

5.35 $n = 10, \bar{d} = 14.89\%$, and $s_d = 30.4868$, with $t_{0.025} = 2.262$ with 9 degrees of freedom. So,

$$14.89 \pm (2.262)\frac{30.4868}{\sqrt{10}} = 14.89 \pm 21.81,$$

which yields $-6.92 < \mu_D < 36.70$.

5.37 $n_A = n_B = 15, \bar{x}_A = 3.82, \bar{x}_B = 4.94, s_A = 0.7794, s_B = 0.7538$, and $s_p = 0.7667$ with $t_{0.025} = 2.048$ with 28 degrees of freedom. So,

$$(4.94 - 3.82) \pm (2.048)(0.7667)\sqrt{1/15 + 1/15} = 1.12 \pm 0.57,$$

which yields $0.55 < \mu_B - \mu_A < 1.69$.

5.39 $n = 1000, \hat{p} = \frac{228}{1000} = 0.228, \hat{q} = 0.772$, and $z_{0.005} = 2.575$. So, using method 1,

$$0.228 \pm (2.575)\sqrt{\frac{(0.228)(0.772)}{1000}} = 0.228 \pm 0.034,$$

which yields $0.194 < p < 0.262$.

When we use method 2, we have

$$\frac{0.228 + 2.575^2/2000}{1 + 2.575^2/1000} \pm \frac{2.575}{1 + 2.575^2/1000}\sqrt{\frac{(0.228)(0.772)}{1000} + \frac{2.575^2}{4(1000)^2}} = 0.2298 \pm 0.0341,$$

which yields an interval of $(0.1957, 0.2639)$.

5.41 (a) $n = 200, \hat{p} = 0.57, \hat{q} = 0.43$, and $z_{0.02} = 2.05$. So,

$$0.57 \pm (2.05)\sqrt{\frac{(0.57)(0.43)}{200}} = 0.57 \pm 0.072,$$

which yields $0.498 < p < 0.642$.

(b) Error $\leq (2.05)\sqrt{\frac{(0.57)(0.43)}{200}} = 0.072$.

5.43 (a) $n = 40, \hat{p} = \frac{34}{40} = 0.85, \hat{q} = 0.15$, and $z_{0.025} = 1.96$. So,

$$0.85 \pm (1.96)\sqrt{\frac{(0.85)(0.15)}{40}} = 0.85 \pm 0.111,$$

which yields $0.739 < p < 0.961$.

(b) Since $p = 0.8$ falls in the confidence interval, we can not conclude that the new system is better.

5.45 $n = \frac{(2.05)^2(0.57)(0.43)}{(0.02)^2} = 2576$ when round up.

5.47 $n = \frac{(2.33)^2(0.08)(0.92)}{(0.05)^2} = 160$ when round up.

5.49 $n = \frac{(1.96)^2}{(4)(0.04)^2} = 601$ when round up.

5.51 $n_M = n_F = 1000, \hat{p}_M = 0.250, \hat{q}_M = 0.750, \hat{p}_F = 0.275, \hat{q}_F = 0.725$, and $z_{0.025} = 1.96$. So

$$(0.275 - 0.250) \pm (1.96)\sqrt{\frac{(0.250)(0.750)}{1000} + \frac{(0.275)(0.725)}{1000}} = 0.025 \pm 0.039,$$

which yields $-0.0136 < p_F - p_M < 0.0636$.

5.53 $n_1 = n_2 = 500, \hat{p}_1 = \frac{120}{500} = 0.24, \hat{p}_2 = \frac{98}{500} = 0.196$, and $z_{0.05} = 1.645$. So,

$$(0.24 - 0.196) \pm (1.645)\sqrt{\frac{(0.24)(0.76)}{500} + \frac{(0.196)(0.804)}{500}} = 0.044 \pm 0.0429,$$

which yields $0.0011 < p_1 - p_2 < 0.0869$. Since 0 is not in this confidence interval, we conclude, at the level of 90% confidence, that inoculation has an effect on the incidence of the disease.

5.55 $s^2 = 0.815$ with $v = 4$ degrees of freedom. Also, $\chi^2_{0.025} = 11.143$ and $\chi^2_{0.975} = 0.484$. So,

$$\frac{(4)(0.815)}{11.143} < \sigma^2 < \frac{(4)(0.815)}{0.484}, \quad \text{which yields} \quad 0.293 < \sigma^2 < 6.736.$$

Since this interval contains 1, the claim that σ^2 seems valid.

5.57 $s^2 = 6.0025$ with $v = 19$ degrees of freedom. Also, $\chi^2_{0.025} = 32.852$ and $\chi^2_{0.975} = 8.907$. Hence,

$$\frac{(19)(6.0025)}{32.852} < \sigma^2 < \frac{(19)(6.0025)}{8.907}, \text{ or } 3.472 < \sigma^2 < 12.804.$$

Chapter 6

One- and Two-Sample Tests of Hypotheses

6.1 (a) Conclude that fewer than 30% of the public are allergic to some cheese products when, in fact, 30% or more are allergic.

(b) Conclude that at least 30% of the public are allergic to some cheese products when, in fact, fewer than 30% are allergic.

6.3 (a) The firm is not guilty.

(b) The firm is guilty.

6.5 (a) $\alpha = P(X \leq 24 \mid p = 0.6) = P(Z < -1.59) = 0.0559.$

(b) $\beta = P(X > 24 \mid p = 0.3) = P(Z > 2.93) = 1 - 0.9983 = 0.0017.$
$\beta = P(X > 24 \mid p = 0.4) = P(Z > 1.30) = 1 - 0.9032 = 0.0968.$
$\beta = P(X > 24 \mid p = 0.5) = P(Z > -0.14) = 1 - 0.4443 = 0.5557.$

6.7 (a) $\alpha = P(X < 110 \mid p = 0.6) + P(X > 130 \mid p = 0.6) = P(Z < -1.52) + P(Z > 1.52) = 2(0.0643) = 0.1286.$

(b) $\beta = P(110 < X < 130 \mid p = 0.5) = P(1.34 < Z < 4.31) = 0.0901.$
$\beta = P(110 < X < 130 \mid p = 0.7) = P(-4.71 < Z < -1.47) = 0.0708.$

(c) The probability of a Type I error is somewhat high for this procedure, although Type II errors are reduced dramatically.

6.9 (a) $n = 12$, $p = 0.7$, and $\alpha = P(X \geq 11) = 0.0712 + 0.0138 = 0.0850.$

(b) $n = 12$, $p = 0.9$, and $\beta = P(X \leq 10) = 0.3410.$

6.11 (a) $n = 70$, $p = 0.4$, $\mu = np = 28$, and $\sigma = \sqrt{npq} = 4.099$, with $z = \frac{23.5-28}{4.099} = -1.10$. Then $\alpha = P(X < 24) = P(Z < -1.10) = 0.1357.$

(b) $n = 70$, $p = 0.3$, $\mu = np = 21$, and $\sigma = \sqrt{npq} = 3.834$, with $z = \frac{23.5-21}{3.834} = 0.65$ Then $\beta = P(X \geq 24) = P(Z > 0.65) = 0.2578.$

6.13 From Exercise 6.12(a) we have $\mu = 240$ and $\sigma = 9.798$. We then obtain

$$z_1 = \frac{214.5 - 240}{9.978} = -2.60, \quad \text{and} \quad z_2 = \frac{265.5 - 240}{9.978} = 2.60.$$

So

$$\alpha = 2P(Z < -2.60) = (2)(0.0047) = 0.0094.$$

Also, from Exercise 6.12(b) we have $\mu = 192$ and $\sigma = 9.992$, with

$$z_1 = \frac{214.5 - 192}{9.992} = 2.25, \quad \text{and} \quad z_2 = \frac{265.5 - 192}{9.992} = 7.36.$$

Therefore,

$$\beta = P(2.25 < Z < 7.36) = 1 - 0.9878 = 0.0122.$$

6.15 (a) $\mu = 200$, $n = 9$, $\sigma = 15$ and $\sigma_{\bar{X}} = \frac{15}{3} = 5$. So,

$$z_1 = \frac{191 - 200}{5} = -1.8, \quad \text{and} \quad z_2 = \frac{209 - 200}{5} = 1.8,$$

with $\alpha = 2P(Z < -1.8) = (2)(0.0359) = 0.0718$.

(b) If $\mu = 215$, then $z_1 = \frac{191-215}{5} = -4.8$ and $z_2 = \frac{209-215}{5} = -1.2$, with

$$\beta = P(-4.8 < Z < -1.2) = 0.1151 - 0 = 0.1151.$$

6.17 (a) $n = 50$, $\mu = 5000$, $\sigma = 120$, and $\sigma_{\bar{X}} = \frac{120}{\sqrt{50}} = 16.971$, with $z = \frac{4970-5000}{16.971} = -1.77$ and $\alpha = P(Z < -1.77) = 0.0384$.

(b) If $\mu = 4970$, then $z = 0$ and hence $\beta = P(Z > 0) = 0.5$.
If $\mu = 4960$, then $z = \frac{4970-4960}{16.971} = 0.59$ and $\beta = P(Z > 0.59) = 0.2776$.

6.19 The hypotheses are

$$H_0 : \mu = 40 \text{ months},$$
$$H_1 : \mu < 40 \text{ months}.$$

Now, $z = \frac{38-40}{5.8/\sqrt{64}} = -2.76$, and P-value$= P(Z < -2.76) = 0.0029$. Decision: reject H_0.

6.21 The hypotheses are

$$H_0 : \mu = 800,$$
$$H_1 : \mu \neq 800.$$

Now, $z = \frac{788-800}{40/\sqrt{30}} = -1.64$, and P-value$= 2P(Z < -1.64) = (2)(0.0505) = 0.1010$. Hence, the mean is not significantly different from 800 for $\alpha < 0.101$.

6.23 The hypotheses are

$$H_0 : \mu = 10,$$
$$H_1 : \mu \neq 10.$$

$\alpha = 0.01$ and $df = 9$.
Critical region: $t < -3.25$ or $t > 3.25$.
Computation: $t = \frac{10.06-10}{0.246/\sqrt{10}} = 0.77$.
Decision: Fail to reject H_0.

6.25 The hypotheses are

$$H_0 : \mu = 20,000 \text{ kilometers},$$
$$H_1 : \mu > 20,000 \text{ kilometers}.$$

Now, $z = \frac{23,500-20,000}{3900/\sqrt{100}} = 8.97$, and P-value$= P(Z > 8.97) \approx 0$. Decision: reject H_0 and conclude that $\mu \neq 20,000$ kilometers.

6.27 The hypotheses are

$$H_0 : \mu_1 = \mu_2,$$
$$H_1 : \mu_1 > \mu_2.$$

Since $s_p = \sqrt{\frac{(29)(10.5)^2+(29)(10.2)^2}{58}} = 10.35$, then

$$P\left[T > \frac{34.0}{10.35\sqrt{1/30 + 1/30}}\right] = P(Z > 12.72) \approx 0.$$

Hence, the conclusion is that running increases the mean RMR in older women.

6.29 The hypotheses are

$$H_0 : \mu = 35 \text{ minutes},$$
$$H_1 : \mu < 35 \text{ minutes}.$$

$\alpha = 0.05$ and $df = 19$.
Critical region: $t < -1.729$.
Computation: $t = \frac{33.1-35}{4.3/\sqrt{20}} = -1.98$.
Decision: Reject H_0 and conclude that it takes less than 35 minutes, on the average, to take the test.

6.31 The hypotheses are

$$H_0 : \mu_A - \mu_B = 12 \text{ kilograms},$$
$$H_1 : \mu_A - \mu_B > 12 \text{ kilograms}.$$

$\alpha = 0.05$.
Critical region: $z > 1.645$.
Computation: $z = \frac{(86.7-77.8)-12}{\sqrt{(6.28)^2/50+(5.61)^2/50}} = -2.60$. So, fail to reject H_0 and conclude that the average tensile strength of thread A does not exceed the average tensile strength of thread B by 12 kilograms.

6.33 The hypotheses are

$$H_0 : \mu_1 - \mu_2 = 0.5 \text{ micromoles per 30 minutes},$$
$$H_1 : \mu_1 - \mu_2 > 0.5 \text{ micromoles per 30 minutes}.$$

$\alpha = 0.01$.
Critical region: $t > 2.485$ with 25 degrees of freedom.
Computation: $s_p^2 = \frac{(14)(1.5)^2+(11)(1.2)^2}{25} = 1.8936$, and $t = \frac{(8.8-7.5)-0.5}{\sqrt{1.8936}\sqrt{1/15+1/12}} = 1.50$. Do not reject H_0.

6.35 Let group 1 is the "no treatment" and group the "treatment." The hypotheses are

$$H_0 : \mu_1 - \mu_2 = 0,$$
$$H_1 : \mu_1 - \mu_2 < 0.$$

$\alpha = 0.05$
Critical region: $t < -1.895$ with 7 degrees of freedom.
Computation: $s_p = \sqrt{\frac{(3)(1.363)+(4)(3.883)}{7}} = 1.674$, and $t = \frac{2.075-2.860}{1.674\sqrt{1/4+1/5}} = -0.70$.
Decision: Do not reject H_0.

6.37 The hypotheses are

$$H_0 : \mu_1 - \mu_2 = 4 \text{ kilometers},$$
$$H_1 : \mu_1 - \mu_2 \neq 4 \text{ kilometers}.$$

$\alpha = 0.10$ and the critical regions are $t < -1.725$ or $t > 1.725$ with 20 degrees of freedom.
Computation: $t = \frac{(16-11)-4}{(0.915)\sqrt{1/12+1/10}} = 2.55$.
Decision: Reject H_0.

6.39 The hypotheses are

$$H_0 : \mu_{II} - \mu_I = 10,$$
$$H_1 : \mu_{II} - \mu_I > 10.$$

$\alpha = 0.1$.
Degrees of freedom is calculated as

$$v = \frac{(78.8/5 + 913.333/7)^2}{(78.8/5)^2/4 + (913/333/7)^2/6} = 7.38,$$

hence we use 7 degrees of freedom with the critical region $t > 1.415$.
Computation: $t = \frac{(110-97.4)-10}{\sqrt{78.800/5+913.333/7}} = 0.22$.
Decision: Fail to reject H_0.

6.41 The hypotheses are

$$H_0 : \mu_1 = \mu_2,$$
$$H_1 : \mu_1 \neq \mu_2.$$

$\alpha = 0.05$.
Degrees of freedom is calculated as

$$v = \frac{(7874.329^2/16 + 2479.503^2/12)^2}{(7874.329^2/16)^2/15 + (2479.503^2/12)^2/11} = 19 \text{ degrees of freedom.}$$

Critical regions $t < -2.093$ or $t > 2.093$.
Computation: $t = \frac{9897.500-4120.833}{\sqrt{7874.329^2/16+2479.503^2/12}} = 2.76$.
Decision: Reject H_0 and conclude that $\mu_1 > \mu_2$.

6.43 The hypotheses are

$$H_0 : \mu_1 = \mu_2,$$
$$H_1 : \mu_1 < \mu_2.$$

Computation: $\bar{d} = -54.13$, $s_d = 83.002$, $t = \frac{-54.13}{83.002/\sqrt{15}} = -2.53$, and $0.01 < P\text{-value} < 0.015$ with 14 degrees of freedom.
Decision: Reject H_0.

6.45 The hypotheses are

$$H_0 : \mu_1 = \mu_2,$$
$$H_1 : \mu_1 > \mu_2.$$

Computation: $\bar{d} = 0.1417$, $s_d = 0.198$, $t = \frac{0.1417}{0.198/\sqrt{12}} = 2.48$ and $0.015 < P\text{-value} < 0.02$ with 11 degrees of freedom.
Decision: Reject H_0 when a significance level is above 0.02.

6.47 $n = \frac{(1.645+1.282)^2(0.24)^2}{0.3^2} = 5.48$. The sample size needed is 6.

6.49 $1 - \beta = 0.95$ so $\beta = 0.05$, $\delta = 3.1$ and $z_{\alpha/2} = z_{0.01} = 2.33$. Therefore,

$$n = \frac{(1.645 + 2.33)^2(6.9)^2}{3.1^2} = 78.28 \approx 79 \text{ due to round up.}$$

6.51 To calculate the sample size, we have

$$n = \frac{(1.645 + 0.842)^2(2.25)^2}{[(1.2)(2.25)]^2} = 4.29.$$

The sample size would be 5.

6.53 (a) The hypotheses are

$$H_0: M_{\text{hot}} - M_{\text{cold}} = 0,$$
$$H_1: M_{\text{hot}} - M_{\text{cold}} \neq 0.$$

(b) Use paired T-test we have $\bar{d} = 726$, $s_d = 207$ with $df = -7$. We obtain $t = 0.99$ with $0.3 <$ 2-sided P-value < 0.4. Hence, fail to reject H_0.

6.55 The hypotheses are

$$H_0: p = 0.40,$$
$$H_1: p > 0.40.$$

Denote by X for those who choose lasagna.

$$P\text{-value} = P(X \geq 9 \mid p = 0.40) = 0.4044.$$

The claim that $p = 0.40$ is not refuted.

6.57 The hypotheses are

$$H_0: p = 0.8,$$
$$H_1: p > 0.8.$$

$\alpha = 0.04$.
Critical region: $z > 1.75$.
Computation: $z = \frac{250-(300)(0.8)}{\sqrt{(300)(0.8)(0.2)}} = 1.44$.
Decision: Fail to reject H_0; it cannot conclude that the new missile system is more accurate.

6.59 The hypotheses are

$$H_0: p = 0.2,$$
$$H_1: p < 0.2.$$

Then

$$P\text{-value} \approx P\left(Z < \frac{136 - (1000)(0.2)}{\sqrt{(1000)(0.2)(0.8)}}\right) = P(Z < -5.06) \approx 0.$$

Decision: Reject H_0; less than 1/5 of the homes in the city are heated by oil.

6.61 The hypotheses are

$$H_0: p_1 = p_2,$$
$$H_1: p_1 > p_2.$$

Computation: $\hat{p} = \frac{29+56}{120+280} = 0.2125$, $z = \frac{(29/120)-(56/280))}{\sqrt{(0.2125)(0.7875)(1/120+1/280)}} = 0.93$, with P-value $= P(Z > 0.93) = 0.1762$.
Decision: Fail to reject H_0. There is no significant evidence to conclude that the new medicine is more effective.

6.63 The hypotheses are

$$H_0 : p_1 = p_2,$$
$$H_1 : p_1 \neq p_2.$$

Computation: $\hat{p} = \frac{63+59}{100+125} = 0.5422$, $z = \frac{(63/100)-(59/125)}{\sqrt{(0.5422)(0.4578)(1/100+1/125)}} = 2.36$, with P-value $= 2P(Z > 2.36) = 0.0182$.
Decision: Reject H_0 at level 0.0182. The proportion of urban residents who favor the nuclear plant is larger than the proportion of suburban residents who favor the nuclear plant.

6.65 The hypotheses are

$$H_0 : p_U = p_R,$$
$$H_1 : p_U > p_R.$$

Computation: $\hat{p} = \frac{20+10}{200+150} = 0.085714$, $z = \frac{(20/200)-(10/150)}{\sqrt{(0.085714)(0.914286)(1/200+1/150)}} = 1.10$, with P-value $= P(Z > 1.10) = 0.1357$.
Decision: Fail to reject H_0. It cannot be shown that breast cancer is more prevalent in the urban community.

6.67 The hypotheses are

H_0 : nuts are mixed in the ratio 5:2:2:1,
H_1 : nuts are not mixed in the ratio 5:2:2:1.

$\alpha = 0.05$.
Critical region: $\chi^2 > 7.815$ with 3 degrees of freedom.
Computation:

Observed	269	112	74	45
Expected	250	100	100	50

$$\chi^2 = \frac{(269-250)^2}{250} + \frac{(112-100)^2}{100} + \frac{(74-100)^2}{100} + \frac{(45-50)^2}{50} = 10.14.$$

Decision: Reject H_0; the nuts are not mixed in the ratio 5:2:2:1.

6.69 The hypotheses are

H_0 : die is balanced,
H_1 : die is unbalanced.

$\alpha = 0.01$.
Critical region: $\chi^2 > 15.086$ with 5 degrees of freedom.
Computation: Since $e_i = 30$, for $i = 1, 2, \ldots, 6$, then

$$\chi^2 = \frac{(28-30)^2}{30} + \frac{(36-30)^2}{30} + \cdots + \frac{(23-30)^2}{30} = 4.47.$$

Decision: Fail to reject H_0; the die is balanced.

6.71 The hypotheses are

$$H_0: f(x) = g(x; 1/2) \text{ for } x = 1, 2, \ldots,$$
$$H_1: f(x) \neq g(x; 1/2).$$

$\alpha = 0.05$.
Computation: $g(x; 1/2) = \frac{1}{2^x}$, for $x = 1, 2, \ldots, 7$ and $P(X \geq 8) = \frac{1}{2^7}$. Hence $e_1 = 128$, $e_2 = 64$, $e_3 = 32$, $e_4 = 16$, $e_5 = 8$, $e_6 = 4$, $e_7 = 2$ and $e_8 = 2$. Combining the last three classes together, we obtain

$$\chi^2 = \frac{(136-128)^2}{128} + \frac{(60-64)^2}{64} + \frac{(34-32)^2}{32} + \frac{(12-16)^2}{16} + \frac{(9-8)^2}{8} + \frac{(5-8)^2}{8} = 3.125$$

Critical region: $\chi^2 > 11.070$ with 5 degrees of freedom.
Decision: Fail to reject H_0; $f(x) = g(x; 1/2)$, for $x = 1, 2, \ldots$

6.73 The hypotheses are

H_0 : Distribution of nicotine contents is normal $n(x; 1.8, 0.4)$,

H_1 : Distribution of nicotine contents is not normal.

$\alpha = 0.01$.

For the total of 30 observations, let's use 6 equal probability intervals, which would lead to $o_i = 5$ for $i = 1, \ldots, 6$. Using $x = 1.8 + (0.4)(z_{1-i/6})$ for $i = 1, \ldots, 6$, we have the following table.

Range	e_i	o_i	$(e_i - o_i)^2/e_i$
$(-\infty, 1.41]$	5	11	7.2
(1.41, 1.63]	5	2	1.8
(1.63,1.8]	5	1	3.2
(1.8,1.97]	5	0	5
(1.8,1.97]	5	0	5
(1.97,2.19]	5	1	3.2
$(2.19, \infty)$	5	15	20
Total	30	30	40.4

$\chi^2 = 40.4$ with critical region: $\chi^2 > 15.086$.
Decision: Reject H_0 and conclude that the distribution is different from $n(x; 1.8, 0.4)$.

6.75 The hypotheses are

H_0 : A person's gender and time spent watching television are independent,

H_1 : A person's gender and time spent watching television are not independent.

$\alpha = 0.01$.
Critical region: $\chi^2 > 6.635$ with 1 degrees of freedom.
Computation:

Observed and expected frequencies			
	Male	Female	Total
Over 25 hours	15 (20.5)	29 (23.5)	44
Under 25 hours	27 (21.5)	19 (24.5)	46
Total	42	48	90

$$\chi^2 = \frac{(15-20.5)^2}{20.5} + \frac{(29-23.5)^2}{23.5} + \frac{(27-21.5)^2}{21.5} + \frac{(19-24.5)^2}{24.5} = 5.4.$$

Decision: Fail to reject H_0; a person's gender and time spent watching television are independent.

6.77 The hypotheses are

H_0 : Occurrence of types of crime is independent of city district,

H_1 : Occurrence of types of crime is dependent upon city district.

$\alpha = 0.01$.
Critical region: $\chi^2 > 21.666$ with 9 degrees of freedom.
Computation:

Observed and expected frequencies					
District	Assault	Burglary	Larceny	Homicide	Total
1	162 (186.4)	118 (125.8)	451 (423.5)	18 (13.3)	749
2	310 (380.0)	196 (256.6)	996 (863.4)	25 (27.1)	1527
3	258 (228.7)	193 (154.4)	458 (519.6)	10 (16.3)	919
4	280 (214.9)	175 (145.2)	390 (488.5)	19 (15.3)	864
Total	1010	682	2295	72	4059

$$\chi^2 = \frac{(162-186.4)^2}{186.4} + \frac{(118-125.8)^2}{125.8} + \cdots + \frac{(19-15.3)^2}{15.3} = 124.59.$$

Decision: Reject H_0; occurrence of types of crime is dependent upon city district.

6.79 The hypotheses are

H_0 : Proportions of household within each standard of living category are equal,

H_1 : Proportions of household within each standard of living category are not equal.

$\alpha = 0.05$.
Critical region: $\chi^2 > 12.592$ with 6 degrees of freedom.
Computation:

Observed and expected frequencies				
Period	Somewhat Better	Same	Not as Good	Total
1980: Jan.	72 (66.6)	144 (145.2)	84 (88.2)	300
May.	63 (66.6)	135 (145.2)	102 (88.2)	300
Sept.	47 (44.4)	100 (96.8)	53 (58.8)	200
1981: Jan.	40 (44.4)	105 (96.8)	55 (58.8)	200
Total	222	484	294	1000

$$\chi^2 = \frac{(72 - 66.6)^2}{66.6} + \frac{(144 - 145.2)^2}{145.2} + \cdots + \frac{(55 - 58.8)^2}{58.8} = 5.92.$$

From Table A.5, $\chi^2 = 5.9$ with 6 degrees of freedom has a P-value between 0.3 and 0.5.

Decision: Fail to reject H_0; proportions of household within each standard of living category are equal.

6.81 The hypotheses are

H_0 : The attitudes among the four counties are homogeneous,

H_1 : The attitudes among the four counties are not homogeneous.

Computation:

Observed and expected frequencies					
	County				
Attitude	Craig	Giles	Franklin	Montgomery	Total
Favor	65 (74.5)	66 (55.9)	40 (37.3)	34 (37.3)	205
Oppose	42 (53.5)	30 (40.1)	33 (26.7)	42 (26.7)	147
No Opinion	93 (72.0)	54 (54.0)	27 (36.0)	24 (36.0)	198
Total	200	150	100	100	550

$$\chi^2 = \frac{(65 - 74.5)^2}{74.5} + \frac{(66 - 55.9)^2}{55.9} + \cdots + \frac{(24 - 36.0)^2}{36.0} = 31.17.$$

Since P-value $= P(\chi^2 > 31.17) < 0.001$ with 6 degrees of freedom, we reject H_0 and conclude that the attitudes among the four counties are not homogeneous.

6.83 The hypotheses are

H_0 : Proportions of voters favoring candidate A, candidate B, or undecided are the same for each city,

H_1 : Proportions of voters favoring candidate A, candidate B, or undecided are not the same for each city.

$\alpha = 0.05$.
Critical region: $\chi^2 > 5.991$ with 2 degrees of freedom.
Computation:

Observed and expected frequencies			
	Richmond	Norfolk	Total
Favor A	204 (214.5)	225 (214.5)	429
Favor B	211 (204.5)	198 (204.5)	409
Undecided	85 (81)	77 (81)	162
Total	500	500	1000

$$\chi^2 = \frac{(204 - 214.5)^2}{214.5} + \frac{(225 - 214.5)^2}{214.5} + \cdots + \frac{(77 - 81)^2}{81} = 1.84.$$

Decision: Fail to reject H_0; the proportions of voters favoring candidate A, candidate B, or undecided are not the same for each city.

Chapter 7

Linear Regression

In this chapter, you may find out that many numbers do not really match the numbers in the computing formulas due to rounding issues in parameter estimations. Many of the final result of each question may come from computer output, which was computed with better precision.

7.1 (a) $\sum_i x_i = 778.7$, $\sum_i y_i = 2050.0$, $\sum_i x_i^2 = 26,591.63$, $\sum_i x_i y_i = 65,164.04$, $n = 25$. Therefore,

$$b_1 = \frac{(25)(65,164.04) - (778.7)(2050.0)}{(25)(26,591.63) - (778.7)^2} = 0.5609,$$

$$b_0 = \frac{2050 - (0.5609)(778.7)}{25} = 64.53.$$

(b) Using the equation $\hat{y} = 64.53 + 0.5609x$ with $x = 30$, we find $\hat{y} = 64.53 + (0.5609)(30) = 81.40$.

(c) Residuals appear to be random as desired.

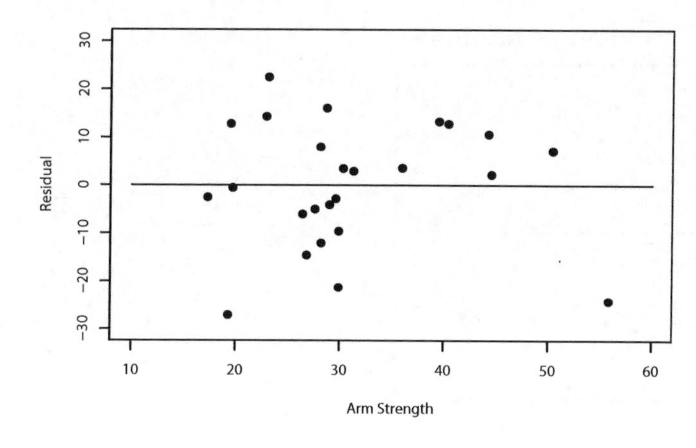

7.3 (a) $\sum_i x_i = 675$, $\sum_i y_i = 488$, $\sum_i x_i^2 = 37,125$, $\sum_i x_i y_i = 25,005$, $n = 18$. Therefore,

$$b_1 = \frac{(18)(25,005) - (675)(488)}{(18)(37,125) - (675)^2} = 0.5676,$$

$$b_0 = \frac{488 - (0.5676)(675)}{18} = 5.8261.$$

Hence $\hat{y} = 5.8261 + 0.5676x$

(b) The scatter plot and the regression line are shown below.

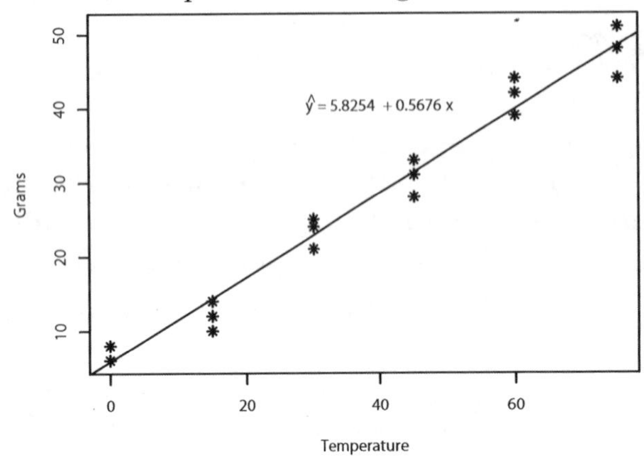

(c) For $x = 50$, $\hat{y} = 5.8261 + (0.5676)(50) = 34.21$ grams.

7.5 (a) $\sum_i x_i = 16.5$, $\sum_i y_i = 100.4$, $\sum_i x_i^2 = 25.85$, $\sum_i x_i y_i = 152.59$, $n = 11$. Therefore,

$$b_1 = \frac{(11)(152.59) - (16.5)(100.4)}{(11)(25.85) - (16.5)^2} = 1.8091,$$

$$b_0 = \frac{100.4 - (1.8091)(16.5)}{11} = 6.4136.$$

Hence $\hat{y} = 6.4136 + 1.8091x$

(b) For $x = 1.75$, $\hat{y} = 6.4136 + (1.8091)(1.75) = 9.580$.

(c) Residuals appear to be random as desired.

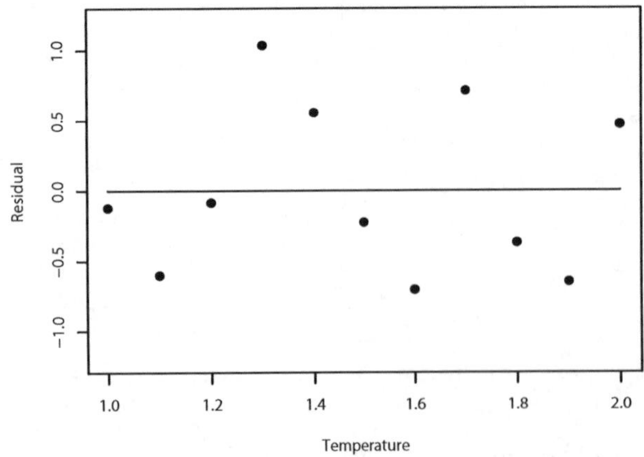

7.7 (a) The scatter plot and the regression line are shown here. A simple linear model seems suitable for the data.

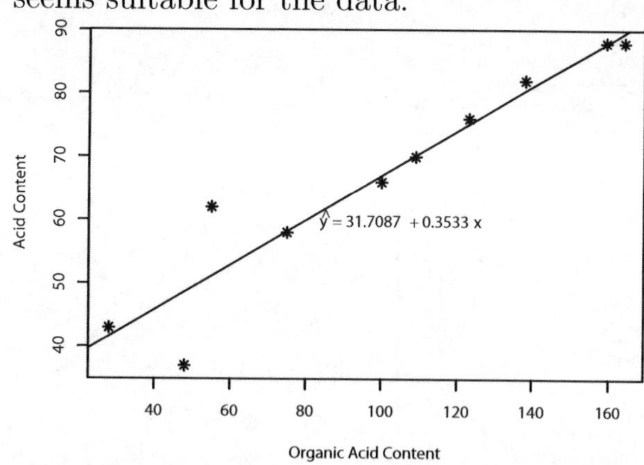

(b) $\sum_i x_i = 999$, $\sum_i y_i = 670$, $\sum_i x_i^2 = 119,969$, $\sum_i x_i y_i = 74,058$, $n = 10$. Therefore,

$$b_1 = \frac{(10)(74,058) - (999)(670)}{(10)(119,969) - (999)^2} = 0.3533,$$

$$b_0 = \frac{670 - (0.3533)(999)}{10} = 31.71.$$

Hence $\hat{y} = 31.71 + 0.3533x$.

(c) See (a).

7.9 (a) The scatter plot and the regression line are shown here.

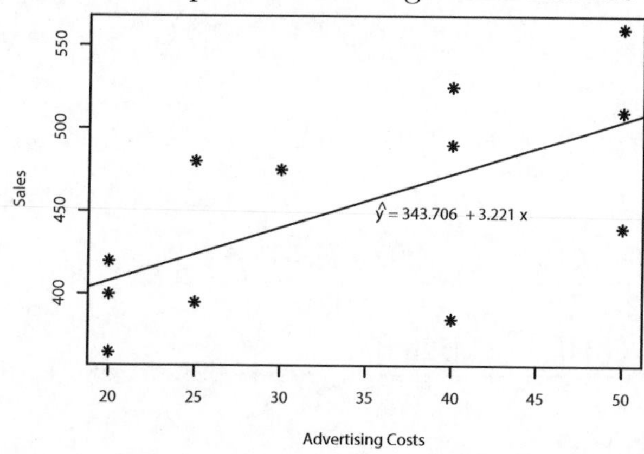

(b) $\sum_i x_i = 410$, $\sum_i y_i = 5445$, $\sum_i x_i^2 = 15,650$, $\sum_i x_i y_i = 191,325$, $n = 12$. Therefore,

$$b_1 = \frac{(12)(191,325) - (410)(5445)}{(12)(15,650) - (410)^2} = 3.2208,$$

$$b_0 = \frac{5445 - (3.2208)(410)}{12} = 343.7056.$$

Hence $\hat{y} = 343.7056 + 3.2208x$

(c) When $x = \$35$, $\hat{y} = 343.7056 + (3.2208)(35) = \$456.43$.

(d) Residuals appear to be random as desired.

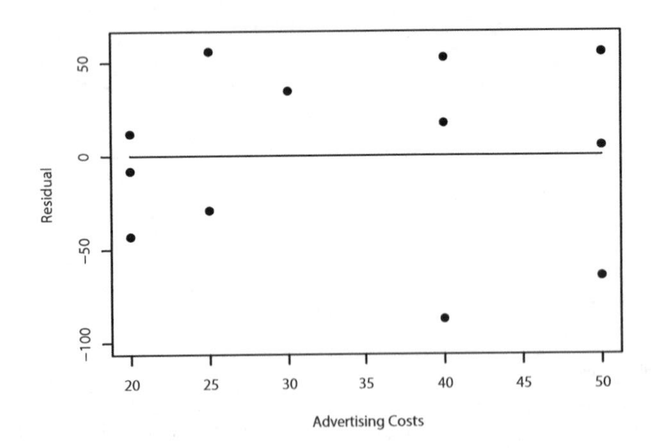

7.11 (a) The scatter plot and the regression line are shown here.

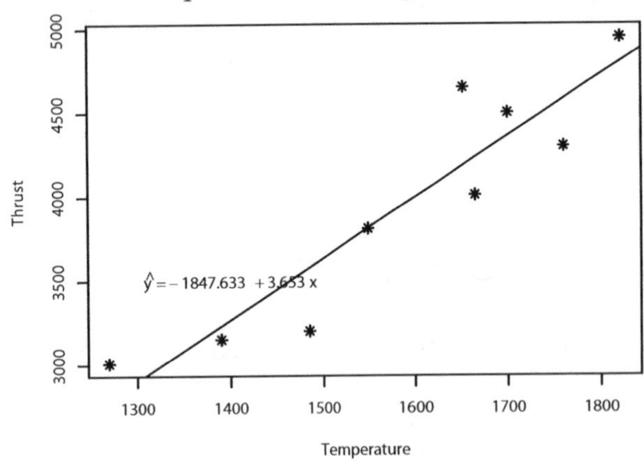

(b) $\sum_i x_i = 14,292$, $\sum_i y_i = 35,578$, $\sum_i x_i^2 = 22,954,054$, $\sum_i x_i y_i = 57,441,610$, $n = 9$. Therefore,

$$b = \frac{(9)(57,441,610) - (14,292)(35,578)}{(9)(22,954,054) - (14,292)^2} = 3.6529,$$

$$a = \frac{35,578 - (3.6529)(14,292)}{9} = -1847.69.$$

Hence $\hat{y} = -1847.69 + 3.6529x$.

7.13 (a) $\sum_i x_i = 45$, $\sum_i y_i = 1094$, $\sum_i x_i^2 = 244.26$, $\sum_i x_i y_i = 5348.2$, $n = 9$.

$$b = \frac{(9)(5348.2) - (45)(1094)}{(9)(244.26) - (45)^2} = -6.3240,$$

$$a = \frac{1094 - (-6.3240)(45)}{9} = 153.1755.$$

Hence $\hat{y} = 153.1755 - 6.3240x$.

(b) For $x = 4.8$, $\hat{y} = 153.1755 - (6.3240)(4.8) = 123$.

7.15 $S_{xx} = 26,591.63 - 778.7^2/25 = 2336.6824$, $S_{yy} = 172,891.46 - 2050^2/25 = 4791.46$, $S_{xy} = 65,164.04 - (778.7)(2050)/25 = 1310.64$, and $b = 0.5609$.

(a) $s^2 = \frac{4791.46-(0.5609)(1310.64)}{23} = 176.362$.

(b) The hypotheses are

$$H_0 : \beta_1 = 0,$$
$$H_1 : \beta_1 \neq 0.$$

$\alpha = 0.05$.
Critical region: $t < -2.069$ or $t > 2.069$.
Computation: $t = \frac{0.5609}{\sqrt{176.362/2336.6824}} = 2.04$.
Decision: Do not reject H_0.

7.17 $S_{xx} = 25.85 - 16.5^2/11 = 1.1$, $S_{yy} = 923.58 - 100.4^2/11 = 7.2018$, $S_{xy} = 152.59 - (165)(100.4)/11 = 1.99$, $b_0 = 6.4136$ and $b_1 = 1.8091$.

(a) $s^2 = \frac{7.2018-(1.8091)(1.99)}{9} = 0.40$.

(b) Since $s = 0.632$ and $t_{0.025} = 2.262$ for 9 degrees of freedom, then a 95% confidence interval is

$$6.4136 \pm (2.262)(0.632)\sqrt{\frac{25.85}{(11)(1.1)}} = 6.4136 \pm 2.0895,$$

which implies $4.324 < \alpha < 8.503$.

(c) $1.8091 \pm (2.262)(0.632)/\sqrt{1.1}$ implies $0.446 < \beta_1 < 3.172$.

7.19 $S_{xx} = 37,125 - 675^2/18 = 11,812.5$, $S_{yy} = 17,142 - 488^2/18 = 3911.7778$, $S_{xy} = 25,005 - (675)(488)/18 = 6705$, $b_0 = 5.8261$ and $b_1 = 0.5676$.

(a) $s^2 = \frac{3911.7778-(0.5676)(6705)}{16} = 6.626$.

(b) Since $s = 2.574$ and $t_{0.005} = 2.921$ for 16 degrees of freedom, then a 99% confidence interval is

$$5.8261 \pm (2.921)(2.574)\sqrt{\frac{37,125}{(18)(11,812.5)}} = 5.8261 \pm 3.1417,$$

which implies $2.686 < \alpha < 8.968$.

(c) $0.5676 \pm (2.921)(2.574)/\sqrt{11,812.5}$ implies $0.498 < \beta_1 < 0.637$.

7.21 The hypotheses are

$$H_0 : \beta_1 = 6,$$
$$H_1 : \beta_1 < 6.$$

$\alpha = 0.025$.
Critical region: $t = -2.228$.
Computations: $S_{xx} = 15,650 - 410^2/12 = 1641.667$, $S_{yy} = 2,512.925 - 5445^2/12 = 42,256.25$, $S_{xy} = 191,325 - (410)(5445)/12 = 5,287.5$, $s^2 = \frac{42,256.25-(3,221)(5,287.5)}{10} = 2,522.521$ and then $s = 50.225$. Now

$$t = \frac{3.221 - 6}{50.225/\sqrt{1641.667}} = -2.24.$$

Decision: Reject H_0 and claim $\beta_1 < 6$.

7.23 Using the value $s = 1.64$ from Exercise 11.20(a) and the fact that $y_0 = 25.7724$ when $x_0 = 24.5$, and $\bar{x} = 25.9667$, we have

(a) $25.7724 \pm (2.228)(1.640)\sqrt{\frac{1}{12} + \frac{(-1.4667)^2}{43.0467}} = 25.7724 \pm 1.3341$ implies $24.438 < \mu_{Y \mid 24.5} < 27.106$.

(b) $25.7724 \pm (2.228)(1.640)\sqrt{1 + \frac{1}{12} + \frac{(-1.4667)^2}{43.0467}} = 25.7724 \pm 3.8898$ implies $21.883 < y_0 < 29.662$.

7.25 Using the value $s = 0.632$ from Exercise 7.17(a) and the fact that $\hat{y}_0 = 9.308$ when $x_0 = 1.6$, and $\bar{x} = 1.5$, we have

$$9.308 \pm (2.262)(0.632)\sqrt{1 + \frac{1}{11} + \frac{0.1^2}{1.1}} = 9.308 \pm 1.4994$$

implies $7.809 < y_0 < 10.807$.

7.27 (a) 17.34.

(b) The goal of 30 mpg is unlikely based on the confidence interval for mean mpg, (26.41, 29.67).

(c) Based on the prediction interval, the mpg of the Lexus ES300 would more likely exceed 18.

7.29 (a) The scatter plot of the data is shown next.

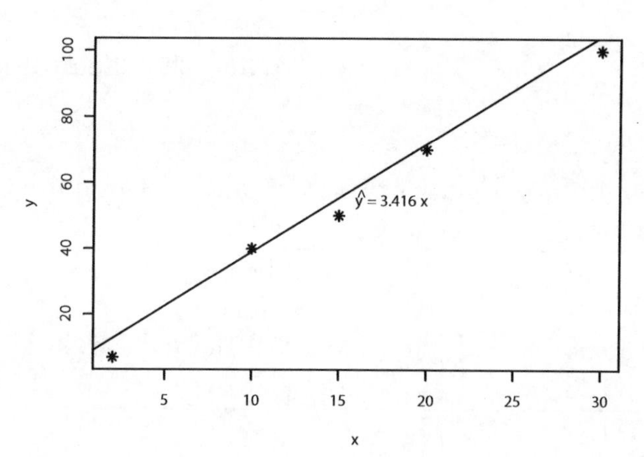

(b) $\sum_{i=1}^{n} x_i^2 = 1629$ and $\sum_{i=1}^{n} x_i y_i = 5564$. Hence $b = \frac{5564}{1629} = 3.4156$. So, $\hat{y} = 3.4156x$.

(c) See (a).

(d) Since there is only one regression coefficient, β, to be estimated, the degrees of freedom in estimating σ^2 is $n - 1$. So,

$$\hat{\sigma}^2 = s^2 = \frac{SSE}{n-1} = \frac{\sum_{i=1}^{n}(y_i - bx_i)^2}{n-1}.$$

(e) $Var(\hat{y}_i) = Var(Bx_i) = x_i^2 Var(B) = \frac{x_i^2\sigma^2}{\sum_{i=1}^{n} x_i^2}$.

(f) The plot is shown next.

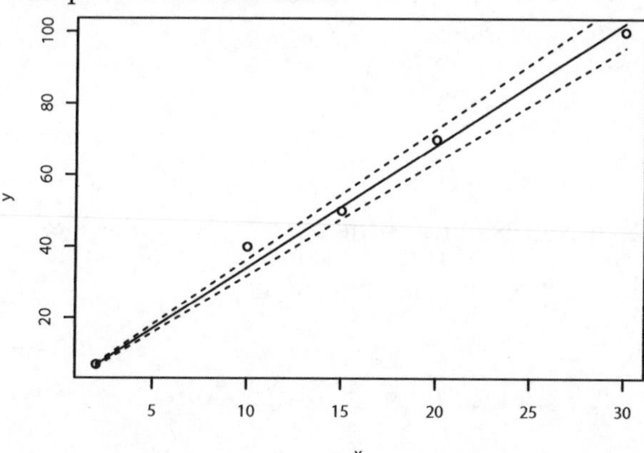

7.31 The hypotheses are

$$H_0: \text{ The regression is linear in } x,$$
$$H_1: \text{ The regression is nonlinear in } x.$$

$\alpha = 0.05$.
Critical regions: $f > 3.26$ with 4 and 12 degrees of freedom.
Computations: $SST = S_{yy} = 3911.78$, $SSR = bS_{xy} = 3805.89$ and $SSE = S_{yy} -$

$SSR = 105.89$. $SSE(\text{pure}) = \sum_{i=1}^{6}\sum_{j=1}^{3} y_{ij}^2 - \sum_{i=1}^{6} \frac{T_{i.}^2}{3} = 69.33$, and the "Lack-of-fit SS" is $105.89 - 69.33 = 36.56$.

Source of Variation	Sum of Squares	Degrees of Freedom	Mean Square	Computed f
Regression	3805.89	1	3805.89	
Error	105.89	16	6.62	
{ Lack of fit	{ 36.56	{ 4	{ 9.14	1.58
{ Pure error	{ 69.33	{ 12	{ 5.78	
Total	3911.78	17		

Decision: Do not reject H_0; the lack-of-fit test is insignificant.

7.33 (a) Since $\sum_{i=1}^{n} x_i y_i = 197.59$, and $\sum_{i=1}^{n} x_i^2 = 98.64$, then $b = \frac{197.59}{98.64} = 2.003$ and $\hat{y} = 2.003x$.

(b) It can be calculated that $b_1 = 1.929$ and $b_0 = 0.349$ and hence $\hat{y} = 0.349 + 1.929x$ when intercept is in the model. To test the hypotheses

$$H_0 : \beta_0 = 0,$$
$$H_1 : \beta_1 \neq 0,$$

with 0.10 level of significance, we have the critical regions as $t < -2.132$ or $t > 2.132$.
Computations: $s^2 = 0.0957$ and $t = \frac{0.349}{\sqrt{(0.0957)(98.64)/(6)(25.14)}} = 1.40$.
Decision: Fail to reject H_0; the intercept appears to be zero.

7.35 (a) $S_{xx} = 1058$, $S_{yy} = 198.76$, $S_{xy} = -363.63$, $b = \frac{S_{xy}}{S_{xx}} = -0.34370$, and $a = \frac{210-(-0.34370)(172.5)}{25} = 10.81153$.

(b) The hypotheses are

$$H_0 : \text{ The regression is linear in } x,$$
$$H_1 : \text{ The regression is nonlinear in } x.$$

$\alpha = 0.05$.
Critical regions: $f > 3.10$ with 3 and 20 degrees of freedom.
Computations: $SST = S_{yy} = 198.76$, $SSR = bS_{xy} = 124.98$ and $SSE = S_{yy} - SSR = 73.98$. Since

$$T_{1.} = 51.1, T_{2.} = 51.5, T_{3.} = 49.3, T_{4.} = 37.0 \text{ and } T_{5.} = 22.1,$$

then

$$SSE(\text{pure}) = \sum_{i=1}^{5}\sum_{j=1}^{5} y_{ij}^2 - \sum_{i=1}^{5} \frac{T_{i.}^2}{5} = 1979.60 - 1910.272 = 69.33.$$

Hence the "Lack-of-fit SS" is $73.78 - 69.33 = 4.45$.

Source of Variation	Sum of Squares	Degrees of Freedom	Mean Square	Computed f
Regression	124.98	1	124.98	
Error	73.98	23	3.22	
{ Lack of fit	{ 4.45	{ 3	{ 1.48	0.43
{ Pure error	{ 69.33	{ 20	{ 3.47	
Total	198.76	24		

Decision: Do not reject H_0.

7.37 $\hat{y} = -21.0280 + 0.4072x$; $f_{\text{LOF}} = 1.71$ with a P-value $= 0.2517$. Hence, lack-of-fit test is insignificant and the linear model is adequate.

7.39 (a) $\hat{y} = -11.3251 - 0.0449$ temperature.

(b) Yes.

(c) 0.9355.

(d) The proportion of impurities does depend on temperature.

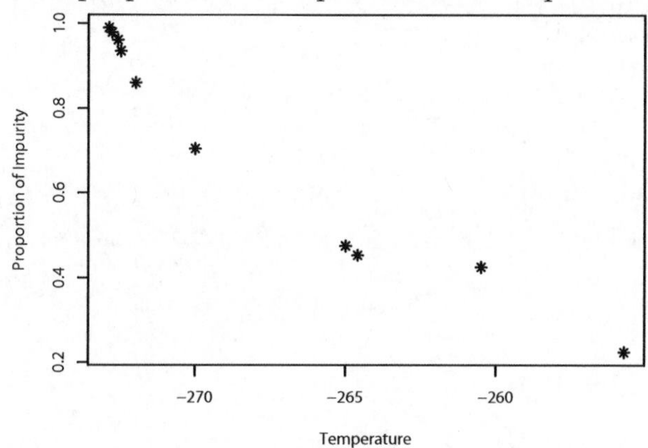

However, based on the plot, it does not appear that the dependence is in linear fashion. If there were replicates, a lack-of-fit test could be performed.

7.41 (a) The figure is shown next.

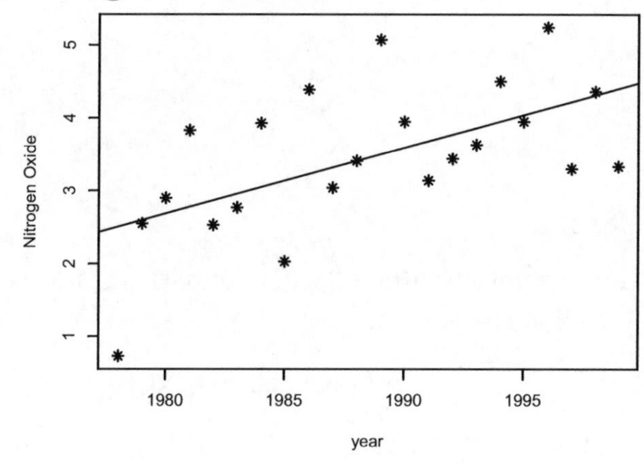

(b) $\hat{y} = -175.9025 + 0.0902$ year; $R^2 = 0.3322$.

(c) There is definitely a relationship between year and nitrogen oxide. It does not appear to be linear.

7.43 $S_{xx} = 36,354 - 35,882.667 = 471.333$, $S_{yy} = 38,254 - 37,762.667 = 491.333$, and $S_{xy} = 36,926 - 36,810.667 = 115.333$. So, $r = \frac{115.333}{\sqrt{(471.333)(491.333)}} = 0.240$.

7.45 (a) From the data of Exercise 7.13 we find $S_{xx} = 244.26 - 45^2/9 = 19.26$, $S_{yy} = 133,786 - 1094^2/9 = 804.2222$, and $S_{xy} = 5348.2 - (45)(1094)/9 = -121.8$. So, $r = \frac{-121.8}{\sqrt{(19.26)(804.2222)}} = -0.979$.

(b) The hypotheses are

$$H_0 : \rho = -0.5,$$
$$H_1 : \rho < -0.5.$$

$\alpha = 0.025$.
Critical regions: $z < -1.96$.
Computations: $z = \frac{\sqrt{6}}{2} \ln\left[\frac{(0.021)(1.5)}{(1.979)(0.5)}\right] = -4.22$.
Decision: Reject H_0: $\rho < -0.5$.

(c) $(-0.979)^2(100\%) = 95.8\%$.

7.47 (a) $S_{xx} = 128.6602 - 32.68^2/9 = 9.9955$, $S_{yy} = 7980.83 - 266.7^2/9 = 77.62$, and $S_{xy} = 990.268 - (32.68)(266.7)/9 = 21.8507$. So, $r = \frac{21.8507}{\sqrt{(9.9955)(77.62)}} = 0.784$.

(b) The hypotheses are

$$H_0 : \rho = 0,$$
$$H_1 : \rho > 0.$$

$\alpha = 0.01$.
Critical regions: $t > 2.998$.
Computations: $t = \frac{0.784\sqrt{7}}{\sqrt{1-0.784^2}} = 3.34$.
Decision: Reject H_0; $\rho > 0$.

(c) $(0.784)^2(100\%) = 61.5\%$.

7.49 $\hat{y} = 0.5800 + 2.7122x_1 + 2.0497x_2$.

7.51 (a) $\widehat{y} = 27.5467 + 0.9217x_1 + 0.2842x_2$.

(b) When $x_1 = 60$ and $x_2 = 4$, the predicted value of the chemistry grade is $\hat{y} = 27.5467 + (0.9217)(60) + (0.2842)(4) = 84$.

7.53 (a) $\hat{y} = -102.71324 + 0.60537x_1 + 8.92364x_2 + 1.43746x_3 + 0.01361x_4$.

(b) $\hat{y} = -102.71324 + (0.60537)(75) + (8.92364)(24) + (1.43746)(90) + (0.01361)(98) = 287.56183$.

7.55 $\hat{y} = 141.61178 - 0.28193x + 0.00031x^2$.

7.57 (a) $\hat{y} = 56.46333 + 0.15253x - 0.00008x^2$.

(b) $\hat{y} = 56.46333 + (0.15253)(225) - (0.00008)(225)^2 = 86.73333\%$.

7.59 $\hat{y} = -6.51221 + 1.99941x_1 - 3.67510x_2 + 2.52449x_3 + 5.15808x_4 + 14.40116x_5$.

7.61 (a) $\hat{y} = 350.99427 - 1.27199x_1 - 0.15390x_2$.

(b) $\hat{y} = 350.99427 - (1.27199)(20) - (0.15390)(1200) = 140.86930$.

7.63 $\hat{y} = 3.3205 + 0.42105x_1 - 0.29578x_2 + 0.01638x_3 + 0.12465x_4$.

7.65 $s^2 = 0.16508$.

7.67 $s^2 = 242.71561$.

7.69 Using *SAS* output, we obtain

(a) $\hat{\sigma}^2_{b_2} = 28.09554$.

(b) $\hat{\sigma}_{b_1 b_4} = -0.00958$.

7.71 The test statistic is $t = \frac{0.00362}{0.000612} = 5.91$ with P-value $= 0.0002$. Reject H_0 and claim that $\beta_1 \neq 0$.

7.73 Using *SAS* output, we obtain
$0.4516 < \mu_{Y|x_1=900,x_2=1} < 1.2083$, and $-0.1640 < y_0 < 1.8239$.

7.75 Using *SAS* output, we obtain
$263.7879 < \mu_{Y|x_1=75,x_2=24,x_3=90,x_4=98} < 311.3357$, and $243.7175 < y_0 < 331.4062$.

7.77 (a) P-value $= 0.3562$. Hence, fail to reject H_0.

(b) P-value $= 0.1841$. Again, fail to reject H_0.

(c) There is not sufficient evidence that the regressors x_1 and x_2 significantly influence the response with the described linear model.

Chapter 8

One-Factor Experiments: General

8.1 The hypotheses are

$$H_0: \mu_1 = \mu_2 = \cdots = \mu_6,$$
$$H_1: \text{At least two of the means are not equal.}$$

$\alpha = 0.05$.
Critical region: $f > 2.77$ with $v_1 = 5$ and $v_2 = 18$ degrees of freedom.
Computation:

Source of Variation	Sum of Squares	Degrees of Freedom	Mean Square	Computed f
Treatment	5.34	5	1.07	0.31
Error	62.64	18	3.48	
Total	67.98	23		

with P-value=0.9024.
Decision: The treatment means do not differ significantly.

8.3 The hypotheses are

$$H_0: \mu_1 = \mu_2 = \mu_3,$$
$$H_1: \text{At least two of the means are not equal.}$$

$\alpha = 0.01$.
Computation:

Source of Variation	Sum of Squares	Degrees of Freedom	Mean Square	Computed f
Shelf Height	399.3	2	199.63	14.52
Error	288.8	21	13.75	
Total	688.0	23		

with P-value=0.0001.
Decision: Reject H_0. The amount of money spent on dog food differs with the shelf height of the display.

8.5 The hypotheses are

$$H_0 : \mu_1 = \mu_2 = \mu_3 = \mu_4,$$
$$H_1 : \text{At least two of the means are not equal.}$$

$\alpha = 0.01$.
Computation:

Source of Variation	Sum of Squares	Degrees of Freedom	Mean Square	Computed f
Treatments	27.5506	3	9.1835	8.38
Error	18.6360	17	1.0962	
Total	46.1865	20		

with P-value= 0.0012.
Decision: Reject H_0. Average specific activities differ.

8.7 The hypotheses are

$$H_0 : \mu_1 = \mu_2 = \mu_3 = \mu_4,$$
$$H_1 : \text{At least two of the means are not equal.}$$

$\alpha = 0.05$.
Computation:

Source of Variation	Sum of Squares	Degrees of Freedom	Mean Square	Computed f
Treatments	119.787	3	39.929	2.25
Error	638.248	36	17.729	
Total	758.035	39		

with P-value=0.0989.
Decision: Fail to reject H_0 at level $\alpha = 0.05$.

8.9 The hypotheses for the Bartlett's test are

$$H_0 : \sigma_1^2 = \sigma_2^2 = \sigma_3^2 = \sigma_4^2,$$
$$H_1 : \text{The variances are not all equal.}$$

$\alpha = 0.01$.
Critical region: We have $n_1 = n_2 = n_3 = 4$, $n_4 = 9$ $N = 21$, and $k = 4$. Therefore, we reject H_0 when

$$b < b_4(0.01, 4, 4, 4, 9)$$
$$= \frac{(4)(0.3475) + (4)(0.3475) + (4)(0.3475) + (9)(0.6892)}{21} = 0.4939.$$

Computation: $s_1^2 = 0.41709$, $s_2^2 = 0.93857$, $s_3^2 = 0.25673$, $s_4^2 = 1.72451$ and hence $s_p^2 = 1.0962$. Therefore,

$$b = \frac{[(0.41709)^3(0.93857)^3(0.25763)^3(1.72451)^8]^{1/17}}{1.0962} = 0.79.$$

Decision: Do not reject H_0; the variances are not significantly different.

8.11 The hypotheses for the Bartlett's test are

$$H_0 : \sigma_1^2 = \sigma_2^2 = \sigma_3^2,$$
$$H_1 : \text{The variances are not all equal.}$$

$\alpha = 0.05$.
Critical region: reject H_0 when

$$b < b_4(0.05, 9, 8, 15) = \frac{(9)(0.7686) + (8)(0.7387) + (15)(0.8632)}{32} = 0.8055.$$

Computation: $b = \frac{[(0.02832)^8(0.16077)^7(0.04310)^{14}]^{1/29}}{0.067426} = 0.7822$.
Decision: Reject H_0; the variances are significantly different.

8.13 (a) The hypotheses are

$$H_0 : \mu_1 = \mu_2 = \mu_3 = \mu_4,$$
$$H_1 : \text{At least two of the means are not equal.}$$

Source of Variation	Sum of Squares	Degrees of Freedom	Mean Square	Computed f
Treatments	1083.60	3	361.20	13.50
Error	1177.68	44	26.77	
Total	2261.28	47		

with P-value< 0.0001.
Decision: Reject H_0. The treatment means are different.

(b) The treatment means are

$$\bar{y}_{1.} = 16.68, \bar{y}_{2.} = 5.24, \bar{y}_{3.} = 17.07, \bar{y}_{4.} = 13.07.$$

Since $q(0.05; 4, 14) \approx 3.79$ from Table A.10, $s^2 = 26.77$ and $n = 12$, the critical difference is 5.66.

Therefore, the results of Tukey's test are

$\bar{y}_{2.}$	$\bar{y}_{4.}$	$\bar{y}_{1.}$	$\bar{y}_{3}$
5.24	13.07	16.68	17.07

8.15 The means of the treatments are:

$$\bar{y}_{1.} = 5.44, \bar{y}_{2.} = 7.90, \bar{y}_{3.} = 4.30, \bar{y}_{4.} = 2.98, \text{ and } \bar{y}_{5.} = 6.96.$$

Since $q(0.05, 5, 20) = 4.24$, the critical difference is $(4.24)\sqrt{\frac{2.9766}{5}} = 3.27$. Therefore, the Tukey's result may be summarized as follows:

$\bar{y}_{4.}$	$\bar{y}_{3.}$	$\bar{y}_{1.}$	$\bar{y}_{5.}$	$\bar{y}_{2.}$
2.98	4.30	5.44	6.96	7.90

8.17 (a) The hypotheses are

$$H_0: \mu_1 = \mu_2 = \cdots = \mu_5,$$
$$H_1: \text{At least two of the means are not equal.}$$

Computation:

Source of Variation	Sum of Squares	Degrees of Freedom	Mean Square	Computed f
Procedures	7828.30	4	1957.08	9.01
Error	3256.50	15	217.10	
Total	11084.80	19		

with P-value= 0.0006.
Decision: Reject H_0. There is a significant difference in the average species count for the different procedures.

(b) Since $q(0.05, 5, 15) = 4.373$ and $\sqrt{\frac{217.10}{4}} = 7.367$, the critical difference is 32.2. Hence

$\bar{y}_K$	$\bar{y}_S$	$\bar{y}_{Sub}$	$\bar{y}_M$	$\bar{y}_D$
12.50	24.25	26.50	55.50	64.25

8.19 The ANOVA table can be obtained as follows:

Source of Variation	Sum of Squares	Degrees of Freedom	Mean Square	Computed f
Temperatures	1268.5333	4	317.1333	70.27
Error	112.8333	25	4.5133	
Total	1381.3667	29		

with P-value< 0.0001.
The results from Tukey's procedure can be obtained as follows:

$\bar{y}_0$	$\bar{y}_{25}$	$\bar{y}_{100}$	$\bar{y}_{75}$	$\bar{y}_{50}$
55.167	60.167	64.167	70.500	72.833

The batteries activated at temperature 50 and 75 have significantly longer activated life.

8.21 Aggregate 4 has a significantly lower absorption rate than the other aggregates.

8.23 The hypotheses are

$$H_0: \alpha_1 = \alpha_2 = \alpha_3 = \alpha_4 = 0, \text{ fertilizer effects are zero}$$
$$H_1: \text{At least one of the } \alpha_i\text{'s is not equal to zero.}$$

$\alpha = 0.05$.
Critical region: $f > 4.76$.
Computation:

Source of Variation	Sum of Squares	Degrees of Freedom	Mean Square	Computed f
Fertilizers	218.1933	3	72.7311	6.11
Blocks	197.6317	2	98.8158	
Error	71.4017	6	11.9003	
Total	487.2267	11		

P-value$= 0.0296$. Decision: Reject H_0. The means are not all equal.

8.25 The hypotheses are

$$H_0: \alpha_1 = \alpha_2 = \alpha_3 = 0, \text{ brand effects are zero}$$
$$H_1: \text{At least one of the } \alpha_i\text{'s is not equal to zero.}$$

$\alpha = 0.05$.
Critical region: $f > 3.84$.
Computation:

Source of Variation	Sum of Squares	Degrees of Freedom	Mean Square	Computed f
Treatments	27.797	2	13.899	5.99
Blocks	16.536	4	4.134	
Error	18.556	8	2.320	
Total	62.889	14		

P-value=0.0257. Decision: Reject H_0; mean percent of foreign additives is not the same for all three brand of jam. The means are:

$$\text{Jam } A\text{: } 2.36, \qquad \text{Jam } B\text{: } 3.48, \qquad \text{Jam } C\text{: } 5.64.$$

Based on the means, Jam A appears to have the smallest amount of foreign additives.

8.27 The hypotheses are

$$H_0 : \alpha_1 = \alpha_2 = \cdots = \alpha_6 = 0, \text{ station effects are zero}$$
$$H_1 : \text{At least one of the } \alpha_i\text{'s is not equal to zero.}$$

$\alpha = 0.01$.
Computation:

Source of Variation	Sum of Squares	Degrees of Freedom	Mean Square	Computed f
Stations	230.127	5	46.025	26.14
Dates	3.259	5	0.652	
Error	44.018	25	1.761	
Total	277.405	35		

P-value< 0.0001. Decision: Reject H_0; the mean concentration is different at the different stations.

8.29 The hypotheses are

$$H_0 : \alpha_1 = \alpha_2 = \alpha_3 = 0, \text{ diet effects are zero}$$
$$H_1 : \text{At least one of the } \alpha_i\text{'s is not equal to zero.}$$

$\alpha = 0.01$.
Computation:

Source of Variation	Sum of Squares	Degrees of Freedom	Mean Square	Computed f
Diets	4297.000	2	2148.500	11.86
Subjects	6033.333	5	1206.667	
Error	1811.667	10	181.167	
Total	12142.000	17		

P-value$= 0.0023$. Decision: Reject H_0; differences among the diets are significant.

8.31 The hypotheses are

$$H_0 : \alpha_1 = \alpha_2 = \alpha_3 = \alpha_4 = \alpha_5 = 0, \text{ treatment effects are zero}$$
$$H_1 : \text{At least one of the } \alpha_i\text{'s is not equal to zero.}$$

$\alpha = 0.01$.
Computation:

Source of Variation	Sum of Squares	Degrees of Freedom	Mean Square	Computed f
Treatments	79630.133	4	19907.533	0.58
Locations	634334.667	5	126866.933	
Error	689106.667	20	34455.333	
Total	1403071.467	29		

P-value= 0.6821. Decision: Do not reject H_0; the treatment means do not differ significantly.

8.33 The hypotheses are

$$H_0 : \alpha_1 = \alpha_2 = \alpha_3 = 0, \text{ dye effects are zero}$$

$$H_1 : \text{At least one of the } \alpha_i\text{'s is not equal to zero.}$$

$\alpha = 0.05$.
Computation:

Source of Variation	Sum of Squares	Degrees of Freedom	Mean Square	Computed f
Amounts	1238.8825	2	619.4413	122.37
Plants	53.7004	1	53.7004	
Error	101.2433	20	5.0622	
Total	1393.8263	23		

P-value< 0.0001. Decision: Reject H_0; color densities of fabric differ significantly for three levels of dyes.

8.35 (a) The model is $y_{ij} = \mu + \alpha_i + \epsilon_{ij}$, where $\alpha_i \sim n(0, \sigma_\alpha^2)$.

(b) Since $s^2 = 0.02056$ and $s_1^2 = 0.01791$, we have $\hat{\sigma}^2 = 0.02056$ and $\frac{s_1^2 - s^2}{10} = \frac{0.01791 - 0.02056}{10} = -0.00027$, which implies $\hat{\sigma}_\alpha^2 = 0$.

8.37 (a) The hypotheses are

$$H_0 : \sigma_\alpha^2 = 0,$$

$$H_1 : \sigma_\alpha^2 \neq 0$$

$\alpha = 0.05$.
Computation:

Source of Variation	Sum of Squares	Degrees of Freedom	Mean Square	Computed f
Operators	371.8719	3	123.9573	14.91
Error	99.7925	12	8.3160	
Total	471.6644	15		

P-value= 0.0002. Decision: Reject H_0; operators are different.

(b) $\hat{\sigma}^2 = 8.316$ and $\hat{\sigma}_\alpha^2 = \frac{123.9573 - 8.3160}{4} = 28.910$.

Chapter 9

Factorial Experiments (Two or More Factors)

9.1 The hypotheses of the three parts are,

(a) for the main effects temperature,

$$H_0^{'}: \alpha_1 = \alpha_2 = \alpha_3 = 0,$$
$$H_1^{'}: \text{At least one of the } \alpha_i\text{'s is not zero;}$$

(b) for the main effects ovens,

$$H_0^{''}: \beta_1 = \beta_2 = \beta_3 = \beta_4 = 0,$$
$$H_1^{''}: \text{At least one of the } \beta_i\text{'s is not zero;}$$

(c) and for the interactions,

$$H_0^{'''}: (\alpha\beta)_{11} = (\alpha\beta)_{12} = \cdots = (\alpha\beta)_{34} = 0,$$
$$H_1^{'''}: \text{At least one of the } (\alpha\beta)_{ij}\text{'s is not zero.}$$

$\alpha = 0.05$.
Critical regions: (a) $f_1 > 3.00$; (b) $f_2 > 3.89$; and (c) $f_3 > 3.49$.
Computations: From the computer printout we have

Source of Variation	Sum of Squares	Degrees of Freedom	Mean Square	Computed f
Temperatures	5194.08	2	2597.0400	8.13
Ovens	4963.12	3	1654.3733	5.18
Interaction	3126.26	6	521.0433	1.63
Error	3833.50	12	319.4583	
Total	17,116.96	23		

Decision: (a) Reject H_0'; (b) Reject H_0''; (c) Do not reject H_0'''.

9.3 The hypotheses of the three parts are,

(a) for the main effects environments,

$$H_0' : \alpha_1 = \alpha_2 = 0, \text{ (no differences in the environment)}$$
$$H_1' : \text{At least one of the } \alpha_i\text{'s is not zero;}$$

(b) for the main effects strains,

$$H_0'' : \beta_1 = \beta_2 = \beta_3 = 0, \text{ (no differences in the strains)}$$
$$H_1'' : \text{At least one of the } \beta_i\text{'s is not zero;}$$

(c) and for the interactions,

$$H_0''' : (\alpha\beta)_{11} = (\alpha\beta)_{12} = \cdots = (\alpha\beta)_{23} = 0, \text{ (environments and strains do not interact)}$$
$$H_1''' : \text{At least one of the } (\alpha\beta)_{ij}\text{'s is not zero.}$$

$\alpha = 0.01$.
Critical regions: (a) $f_1 > 7.29$; (b) $f_2 > 5.16$; and (c) $f_3 > 5.16$.
Computations: From the computer printout we have

Source of Variation	Sum of Squares	Degrees of Freedom	Mean Square	Computed f
Environments	14,875.521	1	14,875.521	14.81
Strains	18,154.167	2	9,077.083	9.04
Interaction	1,235.167	2	617.583	0.61
Error	42,192.625	42	1004.586	
Total	76,457.479	47		

Decision: (a) Reject H_0'; (b) Reject H_0''; (c) Do not reject H_0'''. Interaction is not significant, while both main effects, environment and strain, are all significant.

9.5 The hypotheses of the three parts are,

(a) for the main effects subjects,

$$H_0' : \alpha_1 = \alpha_2 = \alpha_3 = 0,$$
$$H_1' : \text{At least one of the } \alpha_i\text{'s is not zero;}$$

(b) for the main effects muscles,

$$H_0'' : \beta_1 = \beta_2 = \beta_3 = \beta_4 = \beta_5 = 0,$$
$$H_1'' : \text{At least one of the } \beta_i\text{'s is not zero;}$$

(c) and for the interactions,

$$H_0''' : (\alpha\beta)_{11} = (\alpha\beta)_{12} = \cdots = (\alpha\beta)_{35} = 0,$$
$$H_1''' : \text{At least one of the } (\alpha\beta)_{ij}\text{'s is not zero.}$$

$\alpha = 0.01$.
Critical regions: (a) $f_1 > 5.39$; (b) $f_2 > 4.02$; and (c) $f_3 > 3.17$.
Computations: From the computer printout we have

Source of Variation	Sum of Squares	Degrees of Freedom	Mean Square	Computed f
Subjects	4,814.74	2	2,407.37	34.40
Muscles	7,543.87	4	1,885.97	26.95
Interaction	11,362.20	8	1,420.28	20.30
Error	2,099.17	30	69.97	
Total	25,819.98	44		

Decision: (a) Reject H_0'; (b) Reject H_0''; (c) Reject H_0'''.

9.7 The ANOVA table is

Source of Variation	Sum of Squares	Degrees of Freedom	Mean Square	Computed f	P-value
Temperature	430.475	3	143.492	10.85	0.0002
Catalyst	2,466.650	4	616.663	46.63	< 0.0001
Interaction	326.150	12	27.179	2.06	0.0745
Error	264.500	20	13.225		
Total	3,487.775	39			

Decision: All main effects are significant and the interaction is significant at level 0.0745. Hence, if 0.05 significance level is used, interaction is not significant.
An interaction plot is given here.

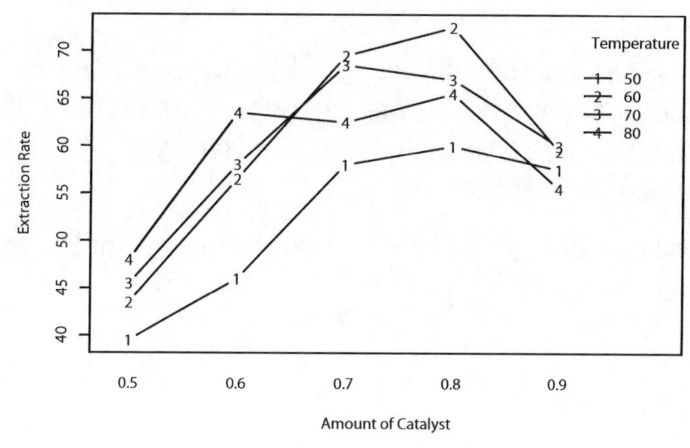

9.9 (a) The ANOVA table is

Source of Variation	Sum of Squares	Degrees of Freedom	Mean Square	Computed f	P-value
Tool	675.00	1	675.00	74.31	< 0.0001
Speed	12.00	1	12.00	1.32	0.2836
Tool*Speed	192.00	1	192.00	21.14	0.0018
Error	72.67	8	9.08		
Total	951.67	11			

Decision: The interaction effects are significant. Although the main effects of speed showed insignificance, we might not make such a conclusion since its effects might be masked by significant interaction.

(b) In the graph shown, we claim that the cutting speed that results in the longest life of the machine tool depends on the tool geometry, although the variability of the life is greater with tool geometry at level 1.

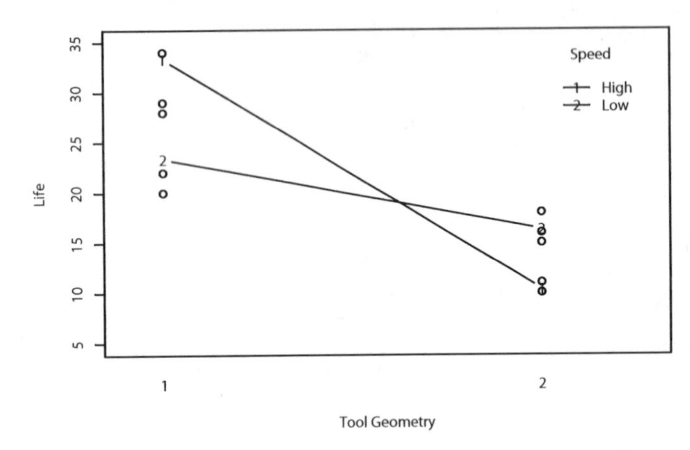

(c) Since interaction effects are significant, we do the analysis of variance for separate tool geometry.

(i) For tool geometry 1, an f-test for the cutting speed resulted in a P-value = 0.0405 with the mean life (standard deviation) of the machine tool at 33.33 (4.04) for high speed and 23.33 (4.16) for low speed. Hence, a high cutting speed has longer life for tool geometry 1.

(ii) For tool geometry 2, an f-test for the cutting speed resulted in a P-value = 0.0031 with the mean life (standard deviation) of the machine tool at 10.33 (0.58) for high speed and 16.33 (1.53) for low speed. Hence, a low cutting speed has longer life for tool geometry 2.

For the above detailed analysis, we note that the standard deviations for the mean life are much higher at tool geometry 1.

(d) See part (b).

9.11 (a) The ANOVA table is

Source of Variation	Sum of Squares	Degrees of Freedom	Mean Square	Computed f	P-value
Method	0.000104	1	0.000104	6.57	0.0226
Lab	0.008058	6	0.001343	84.70	< 0.0001
Method*Lab	0.000198	6	0.000033	2.08	0.1215
Error	0.000222	14	0.000016		
Total	0.00858243	27			

(b) Since the P-value $= 0.1215$ for the interaction, the interaction is not significant. Hence, the results on the main effects can be considered meaningful to the scientist.

(c) Both main effects, method of analysis and laboratory, are all significant.

(d) The interaction plot is show here.

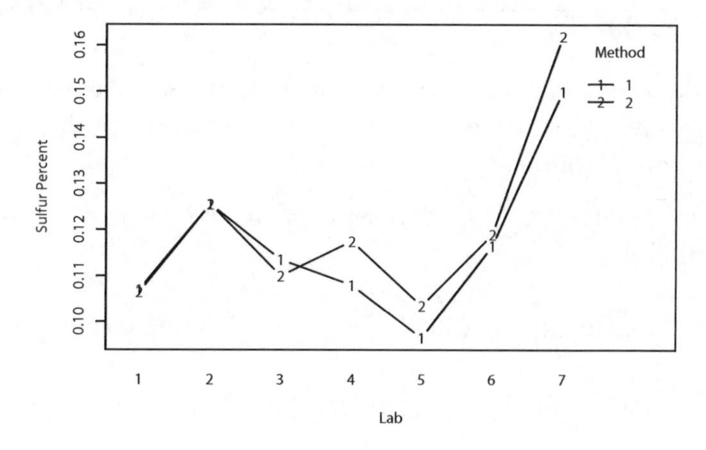

(e) When the tests are done separately, i.e., we only use the data for Lab 1, or Lab 7 alone, the P-values for testing the differences of the methods at Lab 1 and 7 are 0.8600 and 0.1557, respectively. Using partial data such as we did here as above, the degrees of freedom in errors are often smaller (2 in both cases discussed here). Hence, we do not have much power to detect the difference between treatments.

However, if we compare the treatment differences within the full ANOVA model, the degrees of freedom in error can be quite large, e.g., 18 in this case. So, for this case, we obtain the P-values for testing the difference of the methods at Lab 1 and 7 as 0.9010 and 0.0093, respectively. Hence, methods are no difference in Lab 1 and are significantly different in Lab 7. Similar results may be found in the interaction plot in (d).

9.13 (a) The interaction plot is show here. There seems no interaction effect.

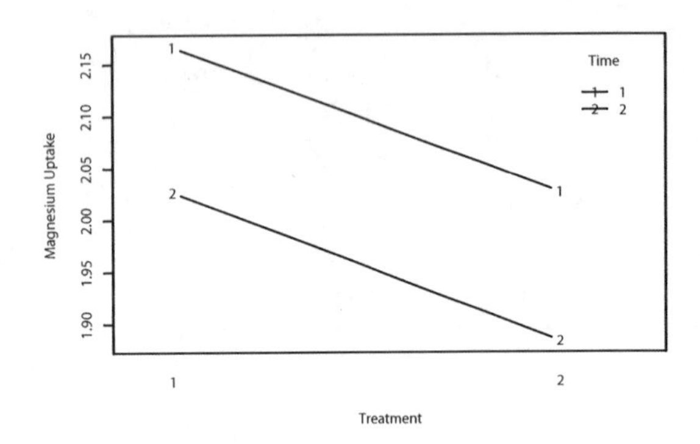

(b) The ANOVA table is

Source of Variation	Sum of Squares	Degrees of Freedom	Mean Square	Computed f	P-value
Treatment	0.060208	1	0.060208	157.07	< 0.0001
Time	0.060208	1	0.060208	157.07	< 0.0001
Treatment*Time	0.000008	1	0.000008	0.02	0.8864
Error	0.003067	8	0.000383		
Total	0.123492	11			

(c) The magnesium uptake are lower using treatment 2 than using treatment 1, no matter what the times are. Also, time 2 has lower magnesium uptake than time 1. All the main effects are significant.

(d) Using the regression model and making "Treatment" as categorical, we have the following fitted model:

$$\hat{y} = 2.4433 - 0.13667 \text{ Treatment} - 0.13667 \text{ Time} - 0.00333 \text{ Treatment} \times \text{Time}.$$

(e) The P-value of the interaction for the above regression model is 0.8864 and hence it is insignificant.

9.15 The ANOVA table is given as follows.

Source of Variation	Sum of Squares	Degrees of Freedom	Mean Square	Computed f	P-value
Launderings	202.13	1	202.13	7.55	0.0087
Bath	715.34	1	715.34	26.73	< 0.0001
Launderings*Bath	166.14	1	166.14	6.21	0.0166
Error	1177.68	44	26.77		
Total	2261.28	47			

(a) The interaction is significant if the level of significance is larger than 0.0166. So, using traditional 0.05 level of significance we would claim that the interaction is significant.

(b) Looking at the ANOVA, it seems that both main effects are significant.

9.17 The ANOVA table is given here.

Source of Variation	Sum of Squares	Degrees of Freedom	Mean Square	Computed f	P-value
Main effect					
A	2.24074	1	2.24074	0.54	0.4652
B	56.31815	2	28.15907	6.85	0.0030
C	17.65148	2	8.82574	3.83	0.1316
Two-factor Interaction					
AB	31.47148	2	15.73574	3.83	0.0311
AC	31.20259	2	15.60130	3.79	0.0320
BC	21.56074	4	5.39019	1.31	0.2845
Three-factor Interaction					
ABC	26.79852	4	6.69963	1.63	0.1881
Error	148.04000	36	4.11221		
Total	335.28370	53			

(a) Based on the P-values, only AB and AC interactions are significant.

(b) The main effect B is significant. However, due to significant interactions mentioned in (a), the insignificance of A and C cannot be counted.

(c) Look at the interaction plot of the mean responses versus C for different cases of A.

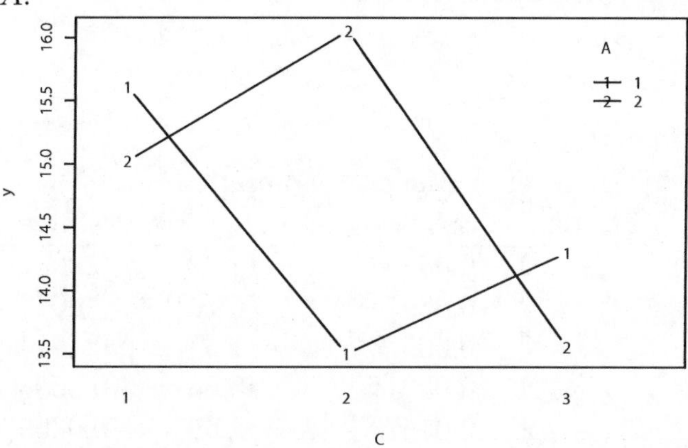

Apparently, the mean responses at different levels of C varies in different patterns for the different levels of A. Hence, although the overall test on factor C is insignificant, it is misleading since the significance of the effect C is masked by the significant interaction between A and C.

9.19 Letting A, B, and C designate coating, humidity, and stress, respectively, the ANOVA table is given here.

Source of Variation	Sum of Squares	Degrees of Freedom	Mean Square	Computed f	P-value
Main effect					
A	216,384.1	1	216,384.1	0.05	0.8299
B	19,876,891.0	2	9,938,445.5	2.13	0.1257
C	427,993,946.4	2	213,996,973.2	45.96	< 0.0001
Two-factor Interaction					
AB	31,736,625	2	15,868,312.9	3.41	0.0385
AC	699,830.1	2	349,915.0	0.08	0.9277
BC	58,623,693.2	4	13,655,923.3	3.15	0.0192
Three-factor Interaction					
ABC	36,034,808.9	4	9,008,702.2	1.93	0.1138
Error	335,213,133.6	72	4,655,738.0		
Total	910,395,313.1	89			

(a) The Coating and Humidity interaction, and the Humidity and Stress interaction have the P-values of 0.0385 and 0.0192, respectively. Hence, they are all significant. On the other hand, the Stress main effect is strongly significant as well. However, both other main effects, Coating and Humidity, cannot be claimed as insignificant, since they are all part of the two significant interactions.

(b) A Stress level of 20 consistently produces low fatigue. It appears to work best with medium humidity and an uncoated surface.

9.21 The ANOVA table shows:

Source of Variation	Sum of Squares	Degrees of Freedom	Mean Square	Computed f	P-value
A	0.16617	2	0.08308	14.22	< 0.0001
B	0.07825	2	0.03913	6.70	0.0020
C	0.01947	2	0.00973	1.67	0.1954
AB	0.12845	4	0.03211	5.50	0.0006
AC	0.06280	4	0.01570	2.69	0.0369
BC	0.12644	4	0.03161	5.41	0.0007
ABC	0.14224	8	0.01765	3.02	0.0051
Error	0.47323	81	0.00584		
Total	1.19603	107			

There is a significant three-way interaction by Temperature, Surface, and Hrc. A plot of each Temperature is given to illustrate the interaction

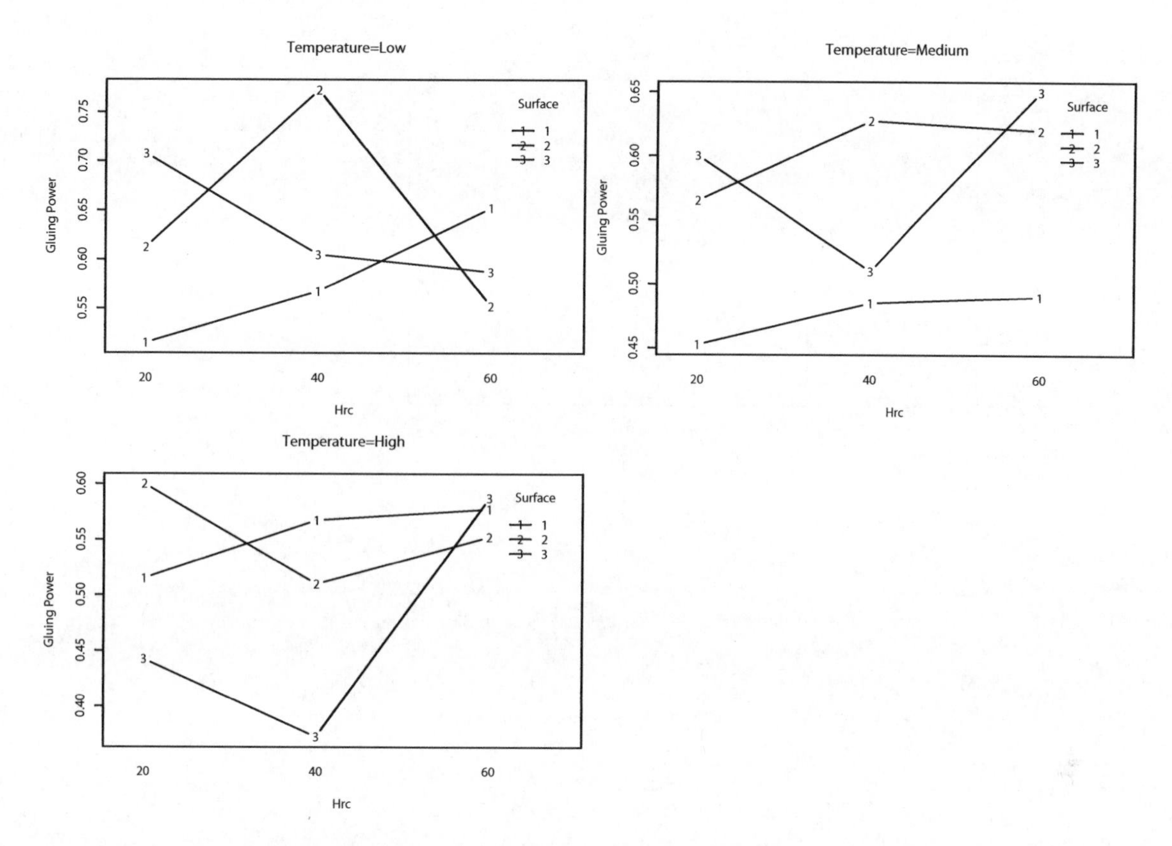

9.23 (a) Yes, the P-values for $Brand * Type$ and $Brand * Temp$ are both < 0.0001.

(b) The main effect of Brand has a P-value < 0.0001. So, three brands averaged across the other two factore are significantly different.

(c) Using brand Y, powdered detergent and hot water yields the highest percent removal of dirt.

9.25 (a) The P-values of two-way interactions Time×Temperature, Time×Solvent, Temperature × Solvent, and the P-value of the three-way interaction Time×Temperature×Solvent are 0.1103, 0.1723, 0.8558, and 0.0140, respectively.

(b) The interaction plots for different levels of Solvent are given here.

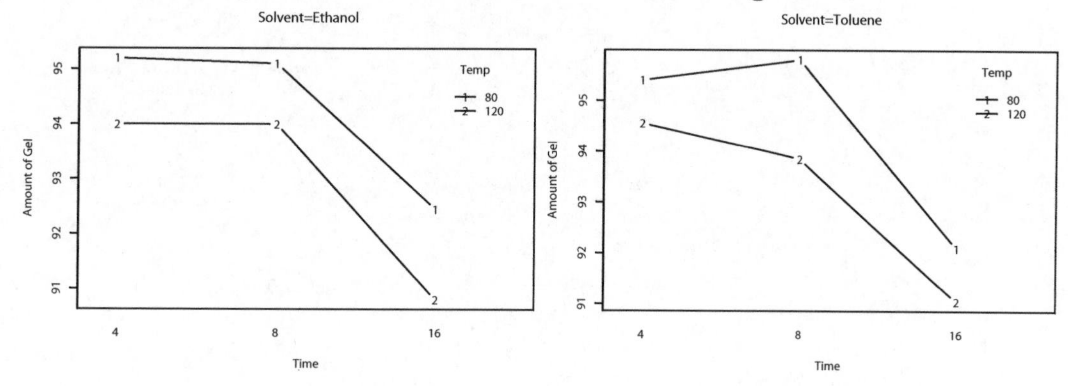

(c) A normal probability plot of the residuals is given and it appears that normality assumption may not be valid.

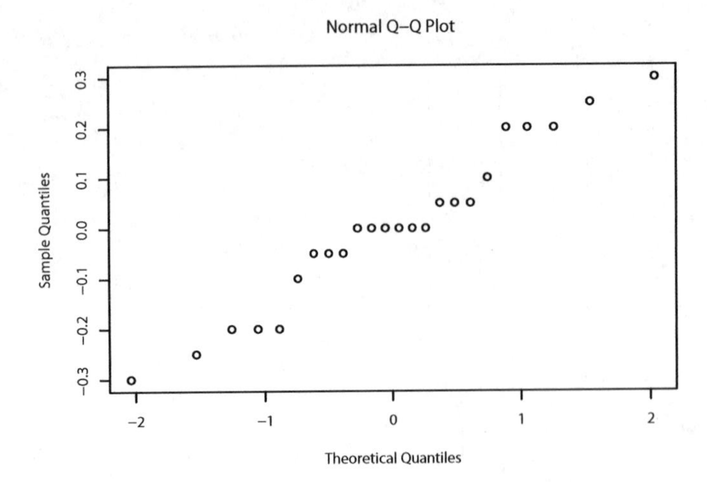
Normal Q–Q Plot
Sample Quantiles
−0.3
−0.2
−0.1
0.0
0.1
0.2
0.3
−2
−1
0
1
2
Theoretical Quantiles